SHIPPING CONTAINER HOMES

THE ULTIMATE BEGINNER'S STEP-BY-STEP GUIDE TO BUILDING YOUR OWN ECO-FRIENDLY AND INEXPENSIVE HOME | WITH PROJECT PLANS, FANTASTIC IDEAS, AND MORE FOR YOUR DREAM HOME.

BY MARK ALLEN

COPYRIGHT

TABLE OF CONTENTS

TABLE OF CONTENTS

TABLE OF CONTENTS

TABLE OF CONTENTS

TABLE OF CONTENTS

TABLE OF CONTENTS

TABLE OF CONTENTS

INTRODUCTION

Why is the interest in transforming shipping containers into homes a growing concept? Though we will get into the benefits and drawbacks in a moment, the main reasons for the growing trend include:

- They're environmentally friendly
- Affordable – nearly half the cost of a traditional home
- It can be customized for any lifestyle

Think shipping container homes is a completely new trend? Not really – early concepts of container living date back to at least 2004 and have picked up steadily since 2010. As the idea of minimizing and going "back to basics" grows worldwide, consumers are looking for more affordable and non-traditional ways to live.

Shopping for a shipping container isn't as easy as one might think, so be ready to dedicate some time and research to find the right option for you and your lifestyle. A wide range of designs and options exist, all of which come with their own respective benefits and drawbacks. Of course, once you purchase your shipping container, there's more work to be done to transform it into a livable home!

In order to understand more about shipping container homes, it's a good idea to describe some of the basic options and design features. Several containers are available, but not all of them are conducive to living – which is why research is important before you get started. Understanding keywords will help the shopping process go much smoother.

Tank containers – These cylindrical containers typically carry liquid materials and are mounted on a steel base.

Dry freight containers – Cubed in shape, these containers open up only with front-facing doors. The opposite end is sealed completely shut.

Open-top containers – Mostly used for materials such as sand or grain, these containers open up from the top and even sides for greater ease in shipping needs.

Thermal containers – Used to ship frozen or temperature-sensitive products or materials, but the container itself provides insulation, no refrigeration.

Refrigeration containers – Commonly known as reefer containers, these shipping containers feature a built-in refrigeration system to keep products or materials cold during transit. Refrigeration containers typically only transport perishable goods.

Global Trends

As one of the more innovative ways to build and design homes, shipping container homes is a growing trend, and there are no signs that such a trend is stopping. Why would people choose to live in a shipping container? There are several reasons, the predominant benefits being that they are easily sustained, affordable, and provide consumers with a modern, compact look.

They work well in high-population urban areas, as well as off-the-grid homes surrounded by limited resources and conveniences! This diversity in housing attracts potential homeowners from all walks of life.

The building time associated with transforming shipping containers into livable homes is also quite rapid in comparison with traditional architecture, making it appealing to those who prefer to move into a home rather quickly.

In addition to being a "trendy" and modern option for eco-friendly consumers, there is also a practical approach to shipping container homes. Homeless and those on a limited income are especially affected positively by the upward development of these homes – the low cost of making and maintaining such homes means that those who don't qualify for traditional homes still have the opportunity to own or rent.

In areas of low income or high demands of housing, shipping container communities provide homeless – including homeless veterans and seniors – with small, welcoming homes or apartments for a very low monthly rent. In the United States, California is one of the first states to develop an apartment building – created using shipping containers – to provide housing for homeless veterans.

These homes are structurally strong and energy-efficient, providing even more benefit to the community as a whole by increasing sustainability and lowering the environmental impact.

Attitudes

Home affordability and living wages are a concern across the globe. The cost of home ownership has also taken on a dramatic climb, making even the smallest home in a less-than-desirable area often too expensive for the average consumer.

So, What's Next?

Shipping containers are the affordable solution to high building costs and housing shortages throughout many of the world's metropolitan areas. These two challenges often come hand-in-hand – the more expensive a market, the harder it is to find necessary housing for those who don't have elevated income levels.

Microdevelopers – or those who go to local communities to purchase vacant land or abandoned properties to develop to sell at a profit – also see a great demand for housing and can make a real dent in supply and demand using affordable, sustainable materials such as shipping containers.

Throughout the world, shipping container communities also have practical uses for young adults. For example, a massive student housing project in Amsterdam has become the largest development for university living. Just as pre-fabricated housing dominated cities after World War II, shipping containers are increasingly becoming more advantageous to overcoming a lack of housing options for people of all ages and economic levels.

How Much Does It Cost?

One of the primary appeals of shipping container living is the affordability factor. Depending on the type of units purchased and the number of units combined to create customized homes, these unique homes can be built for less than $15,000.

Homes that are expanded to multi-levels or installed with luxury living features can range up to $200,000– still much more affordable than a home in countless real estate markets.

It's also important to note that there are a few cost considerations when weighing out the option of building a shipping container home. Depending on the size of the container used, modifications, contractor fees, and materials used throughout – such as flooring or windows – costs vary.

For the sake of comparison, a shipping container home with multiple decks and 2,000 square feet could cost $150,000, while a traditional home of the same size could cost $300,000 on the low end of the scale.

There are many reasons why shipping container homes are a good idea for consumers of all socioeconomic levels and ages. Throughout this book, we will go over some of the important considerations you'll want to make before deciding to move forward. Such as?

Tips for getting started, learning the various design options and how to transform a shipping container into a livable home, common mistakes to avoid, as well as the legal aspect of building shipping container homes, including going over rules and regulations.

CHAPTER 1

Why You Should Build a House of this Type

As the name suggests, a shipping container home is a house that is built from steel containers. It is made from big intermodal containers that help in the transportation of goods and merchandise. Shipping containers can be found in two sizes – 20 feet by 8 feet or 40 feet by 8 feet. Talking about the first one, the 20 feet containers measure around 160 square feet long, and the latter containers are 320 square feet long. You have got the option of using such containers as your home, personal office or just combine many of them to create a multi-story residence. Several shipping container home ideas, like small houses, silo housing, or RVs, are getting more popular day by day. It is because people today want to opt for alternative designs and ideas to the old styles of conventional home buildings.

What Are the Advantages?

When it comes to shipping container homes, there is a wide range of advantages like they are cost-efficient, customizable, and durable.

Cheaper Than Traditional Accommodation

A shipping container can be bought for as little as $10,000. They are quite cost-effective in nature in comparison to conventional housing. It is because labor and construction materials are less needed. You will get the chance to reduce your living costs in a shipping container by using your DIY skills.

Modular In Nature

Container homes can be shipped quite easily and quickly. You can build a house with the use of several containers that range in size from twenty feet to forty feet. You can also include multiple containers for developing a larger house with a dining room, living room, extra floors, and extra bedrooms.

Quick to Construct

It is possible to construct your shipping container home in less than one month. You can also employ a contractor for building your home quickly. But if you decide to opt for a contractor, your cost might rise up for creating your dream home. To cut off high building costs, you will have to opt for a manufactured container home that can be found from various companies that specializes in off-site buildings.

Durable In Nature

Generally, shipping containers are made from Corten steel. Such a kind of self-healing steel helps in supporting cargo when shipping through the water bodies. So, it can be said that homes made from shipping containers will survive better in bad weather compared to standard homes.

Water-Resistant And Watertight

As shipping containers are manufactured with the aim of shipping and transporting goods of all sizes, and for long distances, they are made indestructible. Shipping containers can easily survive heat, earthquakes, water, and even more.

Highly Mobile

You can opt for a specialized shipping provider to collect and relocate a single container house anywhere in the world.

Recycling Is Always Better

Recycled products are one of the best ways of building eco-sustainable homes. You can get expired or "not capable of shipping anymore" containers from anywhere around the world to build your home. At least, reusing another piece of a container is better for the planet.

Eco-Friendly

There are container suppliers who tend to recycle old shipping containers. Based on the user experience, used containers tend to be more conscious environmentally. All such aspects make reusable containers suitable for future container homes.

They Are Stackable

You have got the option to build a small house or a large house for your whole family. You can keep increasing the size of your house by stacking up containers.

Looks Trendy and Cool

You have surely come across images of unusual designs of shipping container homes and hotels. That could be the probable reason why you are here. No matter how you organize and design these large containers, you will always have an esthetic design. If you are looking out for a modern and smart home, this is what you need.

Durable In Nature

Generally, shipping containers are made from Corten steel. Such a kind of self-healing steel helps in supporting cargo when shipping through the water bodies. So, it can be said that homes made from shipping containers will survive better in bad weather compared to standard homes.

Water-Resistant And Watertight

As shipping containers are manufactured with the aim of shipping and transporting goods of all sizes, and for long distances, they are made indestructible. Shipping containers can easily survive heat, earthquakes, water, and even more.

Highly Mobile

You can opt for a specialized shipping provider to collect and relocate a single container house anywhere in the world.

Recycling Is Always Better

Recycled products are one of the best ways of building eco-sustainable homes. You can get expired or "not capable of shipping anymore" containers from anywhere around the world to build your home. At least, reusing another piece of a container is better for the planet.

Eco-Friendly

There are container suppliers who tend to recycle old shipping containers. Based on the user experience, used containers tend to be more conscious environmentally. All such aspects make reusable containers suitable for future container homes.

They Are Stackable

You have got the option to build a small house or a large house for your whole family. You can keep increasing the size of your house by stacking up containers.

Looks Trendy and Cool

You have surely come across images of unusual designs of shipping container homes and hotels. That could be the probable reason why you are here. No matter how you organize and design these large containers, you will always have an esthetic design. If you are looking out for a modern and smart home, this is what you need.

Secure In Nature

Shipping containers are quite tough to break into, and thus, they can make good secure homes. You can definitely install locks, doors, and windows to make your house more secure. But it won't be that easy for burglars in the first place to get into your house.

Are There Any Disadvantages?

Although there is a wide range of advantages when it comes to shipping container homes, there are certain disadvantages as well that you need to pay attention to.

Getting Building Permits Might Not Be That Easy

Getting the license to build your own shipping container home in certain areas is not that easy. But in the US and outside the US, there are certain locations where shipping container homes can be found and are controlled. You will need to consult with the nearest municipal planning office first for more details related to zoning laws, construction codes, and standards of container homes.

Modern Needs Can Be Rigid

If you are not willing to live off the grid, you will have to find an electrician who has the knowledge of building and installing custom electrical systems for homes. Also, you will have to get a plumber to get done with the plumbing work. In case your home location does not have electricity access, you will have to get an installer for adding solar panels on your rooftop.

Need Reinforcements

Although storage containers are made from sturdy steel, alterations, like big windows or doors, might jeopardize the structural integrity. Indeed, shipping containers can easily sustain certain environmental conditions. But in case the corner castings are not that strong, the weight of heavy snow can easily make the roof bow if your location snows. You will have to get in touch with a contractor to reinforce the walls and get a sloping roof.

Small In Size

Shipping containers are not that large in size. Indeed, shipping containers are not that narrow in comparison to small houses on trailers. However, it does not make any kind of huge difference. A shipping container will not be a perfect choice for you if you want complete control over the size of your compact home.

It Can Get Very Hot

You will have to look out for ways in which you can keep the sun off from the roof of your container house as there is a chance of solar heat gain. As they are made of steel, lack of insulation might also make it unbearable to stay in your container home at the time of extreme summer heat.

Shipping Container Home and Its Main Features

Shipping container homes are loaded with features. Such homes make a great option for low-cost, fast, and simple installation. Shipping container homes are prefabricated buildings that are designed originally for some other purpose. However, they are suitable for new structures. Concerning this, one container or more than that can be installed easily and quickly. Constructing a conventional home will eat up a lot of your time, money, and effort. Shipping container homes can be regarded as shortcuts where the preliminary task of building a house is cut off. All that you need to do is to design the interior and exterior a bit. And, if you can work on your own, you can save a lot of your money. They are similar to Lego blocks that can be configured and stacked as you like them.

The houses are made of metal walls and will provide you with all the advantages that a metal structure can provide you with.

- Fire resistant

- No wood rot or mold
- Long-lasting and durable
- Protection from harsh weather

They look modern and stylish. If you are willing to build a house that comes with traditional along with futuristic designs, there is hardly anything else that can beat the look of shipping containers. It is mainly aimed at minimalistic designs and architecture. Shipping containers are eco-friendly in nature. As you transform a container into your home that is completely safe for the environment, you will be upcycling a container. If needed, you can also recycle your house at a later time as a metal frame. You can look out for various designs and ideas online. Or, you can take the help of an architect for the best house designs and transformations.

CHAPTER 2

Chapter 2 Things to Know Before Starting

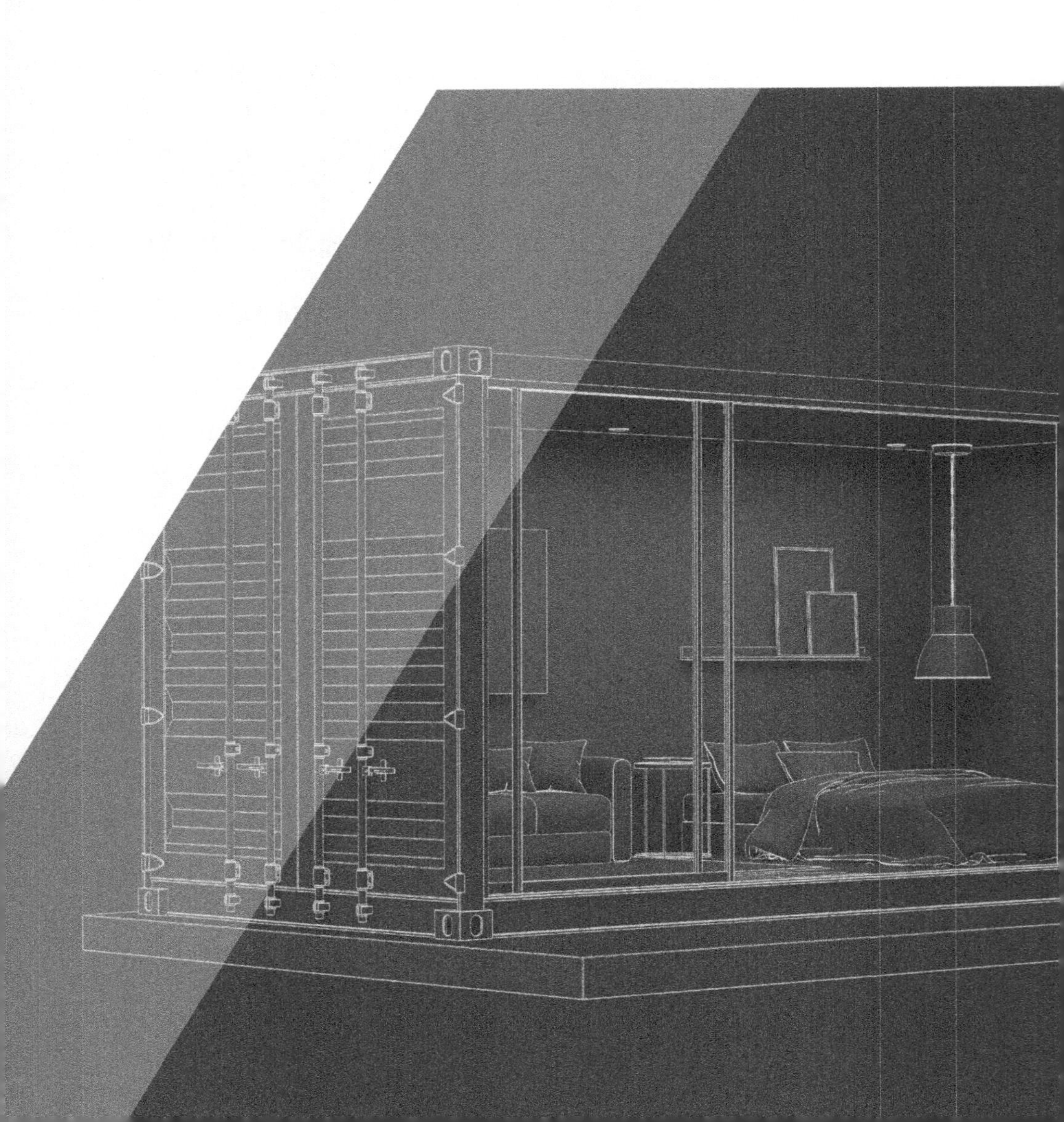

Planning is the real key to a build in an efficient, economical, and achievable way. Even before coming up with a design plan, many factors need to be considered. Proper preparation ensures the construction goes smoothly when it is actually underway.

In the Pre-design stage, the goal is to come up with the concept for the design. The input of those involved in the actual construction, including designers, contractors, and equipment or material suppliers, is taken into account, along with the future home-owner's expectations and the money they are willing to spend.

All in all, this will result in a complete profile of the project: the scope, the budget, and the risks and difficulties to be potentially experienced during the build.

That said, figuring out where to begin can be daunting, especially for a first construction project. Following a checklist (like the one shown here) can be a great way to get started.

Go See an Actual Shipping Container

Shipping containers are needed to build a shipping container home. That is why the first step is to find an actual container and give it a thorough examination. Get a concrete idea of what space looks like. This will help figure out the overall design later and keep expectations in the realm of feasibility.

This is also a chance to compare the quality and price of containers from different sources, should there be more than one in an area. When surveying used containers, keep an eye out for ones that have minimal damage. These may cost a little more, but it is better than going cheap and then overspending on repairs.

Review Building Regulations and Construction Codes that Apply to the Prospective Building Site

A lot will depend on the location of the build. Contact the local building or planning department and inquire about mandatory permits and inspections, as well as any building restrictions. Mention that the home to be constructed has modular steel components and ask if this will lead to any pre-construction issues.

As a rule, most governing bodies grant landowners the right to build almost any residence on privately owned property. However, in some cases, a Certificate of Occupancy may be required. To obtain it, it is necessary to follow building and zoning regulations.

Height limitations, maximum square footage, even the number of bathrooms allowed, are all examples of the kind of information vital in coming up with a final design.

Determine the Budget

It's easy to name a price and call it the budget, but really, that's not how things should be done.

Coming up with a budget requires a look at every aspect of the build. Get an idea of the construction, labor, and professional fees. Take note of everything that needs to be done, what materials will be required and exactly how many people will be needed. Minimize the risk of incurring unexpected expenses by being as thorough as possible.

Consulting architects and contractors is a great way to get a more concrete idea of how much a project will cost. It is also advisable to get in contact with companies who sell containers and do modifications.

It is best to keep the budget a little below the amount of money that is available. This margin will cover any unanticipated expenses that are to be expected during any construction project.

Do a Physical Survey of the Building Site?

The choice of the building site is crucial. One big factor to consider is the soil bearing capacity, referring to the soil's capacity to support the structural load applied to the ground.

Other features of the site, such as the landscape and greenery, can be taken advantage of to provide an aesthetic appeal and a practical one. For example, clumps of trees can naturally provide shade and decrease winter chill when considering the design.

The location of a house will also determine the ease of access and the degree of privacy. Try to see how far the site is from the road or highway. Will it be necessary to add an extended driveway? Concerns like this can increase the cost of the build and should be taken into account during budgeting.

Decide on the Exact Project Requirements

Get started on a "wish list." In particular, focus on the number of rooms that are needed. Decide on how many bedrooms and bathrooms there will be and whether rooms will be constructed for utilities like the kitchen and dining area. Detail any additional features desired, such as a home office space, a game room, or the like.

Estimate the rough square footage required for each, and the whole project.

However, prepare to negotiate. Most likely, adjustments will need to be made in coming up with the final design or the build's course. Try your best to be flexible and to be open to new ideas and suggestions from professionals.

Create a Layout and Floor Plan

Once the "wish list" is ready, start sketching a layout and floor plan for the home. Make sure to draw the dimensions to scale. This will give a better idea of what the finished product will look like after completion.

Remember to incorporate all the functional elements in the home, outlining the correct number of rooms and their purpose. Incorporate as many ideas as possible, but don't overcrowd the space. The key to a comfortable living area is a balanced utilization of space. Even a room with a small area can feel relatively bigger with a successful interior layout.

Finalize the Design

A general idea is all well and good, but once that's done, it's time to get into the specifics. There are a few ways to go about designing a shipping container home.

One is to hire an architect to come up with the design based on the outlined specifications. This has the advantage of producing a unique and original home that will cater to specific needs or preferences.

However, there is a downside. Since shipping container homes are still relatively rare, it may be a challenge to find someone willing and able to design the home. Many design considerations are unique to shipping container homes that the average architect, no matter how skilled, may be unfamiliar with.

An alternative is to locate a design entity that offers turnkey home designs. "Turnkey" means the design is ready to use once purchased. The source company will offer several "stock" designs to choose from. These ready-made "kit" designs can then be adapted to the building site.

Although this doesn't allow for extensive customization, it has the advantage of taking less time and costing less overall. There Should there be non-negotiable aspects of the home, which should be included in the design, look for a company that is willing to customize, although this will probably increase the cost.

Additionally, having the final design is essential in coming up with the final budget. These plans and drawings, among other documents, are also needed in applying for building permits and getting authorization for construction.

CHAPTER 3
How to Select the Container

Like now, there are countless places all through the world from where you can buy a shipping container. Well, as you will notice as you shop around, different sellers have different policies when it comes to inspections and other procedures before the purchase. In essence, try to order containers from local stores as this will allow you to inspect personally as well as avoid the high shipping costs.

If you don't have a dealer within your area, you should consider talking with a member of the ISBU association. This way, you will be assured of the highest level of satisfaction and safety. ISBU (Intermodal Steel Building Units) is the umbrella body for steel building units manufacturers.

However, if you don't want to buy from ISBU, there are many other companies available. You can buy online from companies that deal with shipping tanks.

As you buy the containers, it is paramount that you inspect them first before deciding to buy them. Even if it isn't possible to inspect the container personally, you can do that by checking photos of the container. You can even request other photos from the seller just to be sure about everything.

If you want to buy a used container, you need to ensure it wasn't painted recently. This is because most dealers tend to paint their containers to hide damages. Regardless of its price, please don't take it; if it is nicely painted, you can be sure that there is something that the seller is trying to hide that would warrant painting!

As you have noticed, finding shipping containers isn't just easy. However, just because sellers have shipping containers in good condition doesn't mean that you should just settle on any of them. You should be careful in your selection to be sure that you are choosing the most suitable container.

How to Choose the Best Shipping Containers?

So, you have laid down the plans to build a shipping container home and you have prepared all the other materials but are worried about how to buy the right containers for your project? Well, this shouldn't make you lose sleep.

Before you can even start looking for someone to sell you shipping containers, you should decide whether you want to buy used or new containers. This will make a lot of difference in many things. For instance, it will help you settle on a supplier (some suppliers only sell used containers while others sell new containers), it will help you be mentally prepared in terms of pricing and it will determine the speed with which you make a decision. For instance, when buying a brand-new container, you might not really have to look at so many things like the condition of the container, its age, any history of transporting toxic wastes or materials, and many other things. However, if you opt to buy a used container, you will have to put some essential things into consideration.

Age of the Container

This should give you an estimate of the amount of wear and tear the container has experienced. So even if it has been painted, the years it has been in use can make you think of what to expect. Well, it doesn't mean that a container that has been used for five years will be in a worse condition than that used for say three years. It all depends on usage but given normal usage, you expect the one used for longer to be in a worse condition. You can ascertain the age by checking up the ISO number to determine when it was made.

Dents and Dings

If you are buying a used container, you should expect these. However, this doesn't mean that the structure should be inhabitable because of the dents. As a rule of thumb, if the dents don't damage the structural ability of your building, you can still buy a used container with dents.

Alignment

If a shipping container seems to have lost its original shape, it might not be fit for constructing your house. Well, this isn't always the case but you should be extremely creative to make something that looks appealing.

Rust

Used containers are bound to be rusty as they age. However, this doesn't mean that they should be too rusty otherwise, this would make your house weak.

How Was the Container Used?

People use shipping containers differently; some users even use them to transport highly dangerous chemicals. Your safety comes first even as you shop around for shipping containers to use in constructing your home. If a shipping container had been used to transport hazardous materials whose traces could still remain in the container for years even after cleaning and disinfecting, you shouldn't buy that for home use.

Your goal is to buy a container that will make your life better and easier and not a slow killer!

When you have that in mind, you can move on to finding a good shipping container. Here are some other considerations:

Find and choose a Reputable Distributor

The first thing each potential container homeowner needs to do right is to pick the right distributor. I know you are wondering why then a distributor while manufacturers are known to offer their containers at lower prices. There are many reasons as to this, but the main one is that most manufacturers tend to sell their containers to distributors who order in bulk, which means that you really cannot buy from manufacturers! To be on the safe side, you should check some reviews about the distributor just to be sure of what to expect.

Determine the Best Container Size That Suits Your Needs

The types of containers available have varying sizes; some that are larger than others. Depending on their dimensions, different containers can be used for building homes of different sizes. As such, you should first understand the size of the structure that you want to construct since this will help you settle on an ideal size of the container. Don't just buy extra-large containers that might simply escalate construction costs without offering any utility! Likewise, don't buy very small containers as this may affect your construction plans.

Know the Manufacturer of Such Containers

Knowing who manufactured such containers is very important. Each shipping container manufacturer has a unique selling point in terms of the quality and durability of the containers they make. You might want to know their BBB rating to know people's experiences with containers made by different manufacturers just to give you a rough idea of what to expect.

Having money to buy a shipping container and knowing how to choose a shipping container isn't enough; you have to put into consideration some other points.

CHAPTER 4
Site Preparation

You've purchased your shipping container, and now you're ready for it to be delivered – but as with all things, there are a few considerations you'll want to have in place before the delivery truck shows up.

The first step in preparing the site is exploring the soil type. You'll want to choose the foundation best suited for your site. There are a few options here, and the one you choose will depend upon the soil constitution of your building site.

The most common foundations used for shipping container homes are concrete piers, raft foundations, trench footings, and piles. However, to decide what's best for you, you need to know a bit about the strengths of each and the soil of your location. Some prefer to "over-spec" their foundations, creating them to withstand forces stronger than will ever conceivably be required by the building. It's really up to you once you have explored the options.

Soil Types

Dirt is just dirt, right? Wrong. There are many different soil types, and each holds a structural load differently. Here is a list of the soil types you may encounter in your building site and some of the construction aspects related to them:

Rock

If your building site is a slab of rock, it's actually a blessing. You need very little to build upon this, other than to ensure that it's level and strip the surface soil. This rock can consist of granite, schist, or diabase. It has a high load-bearing capacity and can hold anything you place upon it. One of the cheapest and easiest foundations to use in these conditions is the concrete pier foundation.

Gravel

We probably all know about gravel. It is coarse-grained material, easily dug out, and offering excellent drainage. The best foundation for this type of soil is the trench footing.

Clay

Clay is extremely fine-grained and holds water. If your building site has a clay-based soil, you may be in for some costly preparation. You'll need to dig down until you find more stable soil. For this soil type, pile foundations or deep trench footings are most appropriate.

Sandy Soil

Sand consists of fine-grained particles, often containing a mix of gravel and rock. When working with these soils, it's important not to over-dig the foundation. This can increase the load on the softer soils below the layer of sand. The best foundation for this soil type is the raft foundation, often called a slab-on-grade.

What Soil Type Do I Have?

Unless you're a geotechnical engineer or the soil type is fairly obvious, it's best to seek out a bit of professional help for this stage of the game. The geotechnical engineer will explore the entire site, doing soil borings to evaluate the soil constitution at different depths. This will give you a solid and comprehensive idea of what you're working with and the best foundation to support your dwelling.

Typically, a soil engineer will perform test bores across the site, spaced 100 to 150 ft. apart. These test bearings will supply a soil profile, indicating the load-bearable soil. These tests will indicate the water content, density, and particle size of the soil, as well as the soil classification, depth of different soil types, and groundwater level of your site. Bores will be continued into at least 20 ft. of load-bearing soil.

Geotechnical engineers will also explore the surface qualities of the soil, identifying soils that challenge the building process. They will explore the elevation of the site, offering indications as to how the site can be leveled most effectively. In essence, the geotechnical engineer will provide a report which indicates:

- Surface soil type
- Subsurface soil type
- Soil bearing capacity
- Groundwater depth
- Frost depth
- Soil compaction
- Recommended foundation type

- Recommended foundation depth
- Drainage requirements

Tip: Sometimes, the local authority has information on the soil profile of your area. If you plan to build within city limits, this is a high likelihood. If you are building something a bit further out, you can almost guarantee that a soil profile will be necessary, and should factor this into your budget.

As we've talked about before, before scheduling a delivery for your shipping container, make sure that the land on which the container will be placed is allowed for a residential living (more about permits and zoning laws later).

Once you have all the legal aspects in place, you can move forward with preparing the site for delivery. Here are the various steps to complete so that the delivery of your shipping container may run as smoothly as possible.

Communicate!

Communication with the delivery driver is one of the most important details that you must not overlook. Before the delivery driver shows up with your shipping container, you will both need to have an idea of the logistics of delivering the shipping container to a designated site.

Once the delivery driver knows more about the location and how they can reach the particular site, he or she will be much more able to come up with a plan on how to deliver the container to the precise location. Knowing the area in advance also allows the delivery driver to consider what additional equipment they will need to make the job easier.

The desired location of your doors and windows on the shipping container will also need to be deliberated with the delivery driver. This is a crucial step – how the container is loaded onto the flatbed of the truck will directly impact how it comes off when it's delivered. Once at the site, there is no easy way to turn and rotate the shipping container, so be sure to deliberate this important detail with the delivery driver before they arrive at the loading destination.

Choose a Delivery Method

The most common way for a shipping container to be delivered to a location is by truck, but some options are depending on the size or style of the shipping container. There are two standard trailer options, including:

- Flatbed trailers
- Tilt-bed trailers

Which one of these trailers is a better option? It depends on what other equipment you'll have at the delivery site and your budget.

For example, a tilt-bed trailer is a good option if you do not have a crane or forklift designated for the size and shape of a shipping container.

This trailer bed style allows the delivery driver to drop off the container in the designated location, without requiring additional equipment. There is a consideration to make when it comes to tilt-bed trailers: these only accommodate shipping containers that range between 20 and 40 feet.

Another option is the flatbed trailer, which is the least expensive but requires additional tools, including a forklift or crane, which incurs an additional charge. Once you know where you want to place the shipping container, you can determine if the cost savings when using a flatbed trailer is better than opting for the tilt-bed trailer.

Preparing the Land for a Shipping Container

Before the delivery truck brings your shipping container, the land where it will be placed must be ready for such a container – once the container is delivered, there is no easy way to move it so you can make adjustments!

The ground where you want the shipping container must be completely flat and firm. If it is not, you run the risk of structurally compromising the home's integrity or having an uneven surface – both of which have a huge impact on your shipping container home before it's even placed on the ground.

Another thing to consider is potential flooding – even if you do not plan to place your shipping container near any sources of water, you must always allow for proper drainage of water away from the container. This means never placing the container in a low-lying area – even the slightest elevation surrounding you can create a potential disaster if there is rain or snow.

Proper drainage also impacts the quality of your foundation. If you don't have a way for water to leave the area, you risk it softening up the foundation under your shipping container and causing it to sink into the ground – this will cause your home to be structurally unsafe. Depending on your ground conditions, you can choose from a variety of foundation options:

- Gravel
- Hard grass
- Compact dirt
- Cement
- Pavement

If you do not want to use these options, you can also use concrete blocks/footings for small containers. These blocks/footings are to be placed under the corners of the container, elevating it up from the ground and preventing sinking. This option is not the most ideal, but it provides an alternative if the other foundation options are not available. The table below will give you an estimation of the bearing of each footing.

Minimum Width of Concrete or Masonry Footings (inches)						
	Load-Bearing Value of Soil (psf)					
	1,500	2,000	2,500	3,000	3,500	4,000
Conventional Wood Frame Construction						
1-story	16	12	10	8	7	6
2-story	19	15	12	10	8	7
3-story	22	17	14	11	10	9
4-Inch Brick Veneer Over Wood Frame or 8-Inch Hollow Concrete Masonry						
1-story	19	15	12	10	8	7
2-story	25	19	15	13	11	10
3-story	31	23	19	16	13	12
8-Inch Solid or Fully Grouted Masonry						
1-story	22	17	13	11	10	9
2-story	31	23	19	16	13	12
3-story	40	30	24	20	17	15
Source: Table 403.1; *CABO One- and Two- Family Dwelling Code*; 1995.						

For those who don't mind paying a bit more, a chassis is another option that also works well if you plan to move your shipping container at any point in the future. A chassis is a type of foundation used to secure a shipping container in place during the transportation process.

What to Do on Delivery Day

The delivery day has arrived, and it's an exciting step in the process of building your shipping container home!

There are some logistic considerations to keep in mind

regarding delivery – the most important thing is that the delivery truck and trailer is going to be rather large, so it must be able to reach the delivery location and get out.

Before the delivery truck arrives, be sure to clear away anything that might be in the path of the truck – plants and trees included. Shipping containers weigh tons – so they're not simple to move around! If you have existing landscaping around the delivery site, be prepared for it to experience some markings, especially if you have grass in the area, as the trucks and trailers are heavy as well!

For the best results, the delivery truck driver will need to back up to the designated space, then use a tilt-bed or crane to remove the shipping container from the trailer and place it down onto the foundation.

Dealing With Permits

Permits – it's the word that often brings up feelings of dread when it comes to building a home. In the case of shipping container homes, the concept of permits cannot be avoided. Despite being a substantial obstacle between building and moving into the home, there are a few reasons why permits are important.

- They serve to protect the health of the inhabitants in the home.

- Ensure the home is safe for dwellers and the land around it.

- Keep residential areas consistent and cohesive with the rest of the neighborhood.

As you have garnered by now, shipping container homes are not traditional – which means that permits and codes designated for traditional homes don't always apply to a structure such as a steel shipping container.

The trend of shipping container homes is still new enough that there are very few official regulations on the book regarding how those homes can be built safely and "according to code." This makes securing permits a bit harder and potentially drawn out.

Another thing to consider is that codes differentiate between cities, due to codes being created at a municipal level. This means that codes you might need to follow in Los Angeles are much different from those elsewhere, such as Denver or Salt Lake City.

Since building permits can take some time to secure, it's always a good idea to contact your city government offices for specific information about building codes in your area before getting started. You certainly don't want to run into problems when you're nearing the completion of your shipping container home because you learned that some aspects are not acceptable for living.

Regulations cover several aspects of your home, including:

- Installation

- Doors and windows

- Fire regulations
- Climate control

Another reason for making sure your home follows proper codes is that your home insurance will depend on it – be sure to follow through with an insurance agent to ensure that your home will follow all the necessary requirements for coverage!

Do I Need a Permit?

Every city in the country has a different set of rules and regulations, which means that obtaining a permit will highly depend on what the laws are for your given city. The United States follows the International Building Code (IBC), so every city or jurisdiction will enforce the codes that they feel are necessary for the location.

Building codes are often needed when it comes to adding to or modifying the structural integrity of the home, including:

- Electricity
- Plumbing
- Window modifications
- Mechanical systems
- Sewage modifications

Depending on the design of the home, exterior work might also require a permit – removing retaining walls, adding fencing, or even installing a deck around the home all fall under the possibilities of needing a permit. In any case, always check with the local government office or city hall to determine if your plans require permits before you get started.

When it comes to shipping container homes, there are several reasons why it might be hard to secure a permit. For example, while a steel shipping container is structurally stronger than a traditional house, much of the load-bearing weight is located in its frame, not the walls.

Also, there is no insulation and it needs a complete re-modification to transform from a steel container into a home – walls, floors, ventilation, and electricity all need to be added for it to be livable.

Where To Obtain a Permit

Your best bet for understanding exactly what you need in your location will depend on what your municipality requires – every city is different. Be prepared for a bit of a slowdown, too, as shipping container homes are still very new, and therefore, many municipalities – especially those in more rural areas – still don't have much in the way of permits and codes available for these types of homes.

If you aren't sure where to start – or when to start – it is recommended that you speak to the building department before you begin construction. To help articulate what you plan on doing, have your building designs and some other examples of shipping container homes, just to illustrate your plan to those who might not be familiar. If you can find examples of shipping container homes in your state, even better!

Tips for Dealing with Permits

There's usually no easy way around the concept of permits. These tips will make things a bit easier and hopefully prevent you from making a costly and time-consuming mistake.

- Verify the zoning information for your desired plot of land from the zoning department at City Hall before you start building. This cannot be stressed enough.

- Determine what rules you will need to follow to accommodate zone specifications.

- Prepare and show construction plans to City Hall to determine what permits you'll need.

- Get ready to build!

CHAPTER 5
Planning

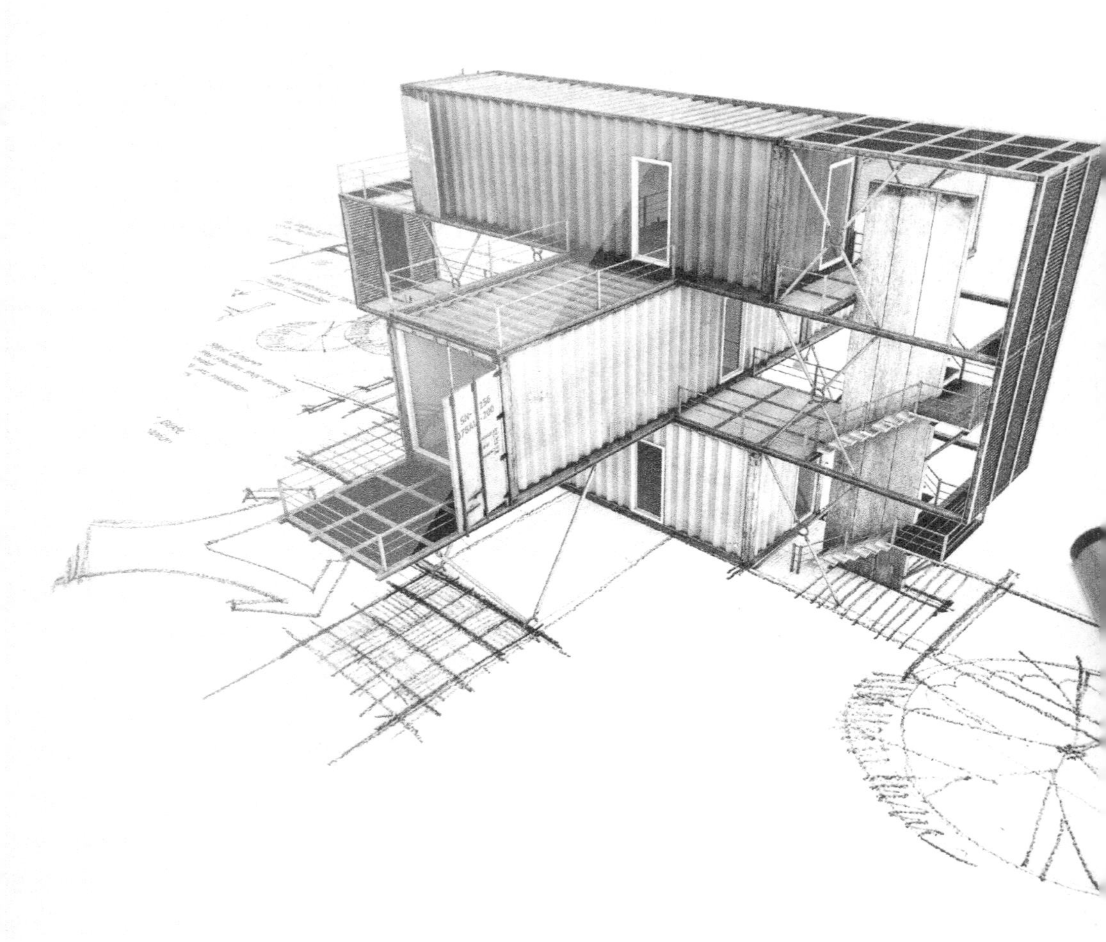

Important Questions to Start With

To know how you'd like to build your shipping container home; the first step is to plan it out. This means asking a few questions and exploring some practical needs and considerations before entering the design phase. To get you started with this, here are some of the most important things to take into consideration:

What Are Your Needs?

Consider how many bedrooms you want, how much floor space you and your family will require.

Consider the number of bathrooms and the extent of storage space necessary.

With a shipping container home, space is at a premium, so you'll want to consider how you can accommodate all of your needs as efficiently as possible.

Where Will the Home Be Built?

This question influences several other considerations, including zoning requirements for your intended location, soil type, and foundation, and utilities planning.

What's Your Budget?

This is another key consideration, as it will determine how many containers you can use, whether or not you will be able to hire contractors to help with the work, whether you intend to buy new, used, or one-time use containers, and several other factors.

When Would You Like to Have Your Home Completed?

Timing is crucial when considering container delivery, equipment rental, and contracting arrangements.

Remember that in most instances, shipping container homes can be built in a matter of days or at most weeks.

Can You Pull It Off?

Do you have access to the skills, materials, financial resources, planning permission, and other necessities required for construction?

Financial Planning

What can you afford? In most instances, this will determine your budget. Key considerations will include the cost of the land, the container, and the roof. Other important considerations are whether you have the skills, time, and resources to do the modification yourself, if you would like to hire contractors for the work, or if you intend to purchase the container pre-modified as a home.

Among these factors are whether you choose to purchase a new, used, or one-time-use container, including a roof, and the degree of modification you would like done to the container. Other important factors to consider are the required foundation type, utilities, and the transport and placement of the container.

Here is the budget used for the real-life example depicted above. This will give you a better idea of some of the possible costs.

Component	Quantity	Expenditure
Land (including taxes and fees)	1	$8000
Fees to professionals		$6500
Standard 40 Foot Container (Used in US)	2	$6300
Bathroom		$2500
Door (for both inside and outside)	5	$700
Kitchen		$3000
Insulation	1	$1800
Furniture		$3800
Windows	4	$850
Utilities		$3500
Internal fixes and fittings		$1500
Transport		$500
Flooring		$2300
Roof		$5500
Paint		$700
Walls inside home		$300
Unexpected costs	12%	$5730
Summary		$53480

One thing to remember regarding the example above is that the furnishings, flooring, and interior design have been chosen with luxury rather than the economy in mind. Note that a 10% contingency cost has been added. Most construction endeavors, regardless of how well planned, include some unexpected expenses, so it's helpful to make sure you're prepared for such things when they occur.

Permits and Zoning Laws

Each country and, in most cases, each zoning district within a country has its own zoning regulations. These regulations determine which types of buildings can be placed on a given lot, as well as the density, height, and other requirements for structures within the zone. These regulations are complex, and it is not possible to provide all the details for every zone. However, here are a few details that will give some direction in becoming knowledgeable about your country's guidelines. Also included is a list of documents that are likely to be required by any country when designing your home.

Australia

Before major building work in Australia, you must obtain a permit from the local council. Check into your state's policy planning framework online, and then approach the council to find out the requirements for your state and for the council which governs your intended building site. They will be able to provide you a list of any documents required for your area or any regulations which should be considered during the design and planning phases.

New Zealand

New Zealand is a bit ahead of the game when it comes to shipping container homes. The Building Act of 2004 offers clear guidance for the construction of these homes. In most cases, they will require building consent, though if they are intended only for storage, then they may be exempt from this requirement. In addition, the territorial authority may choose to exempt your shipping container home from building consent, so long as it still meets the building codes. As in the examples above, the first step is to confer with the territorial authority to look into the specifics of your intended site and design.

United Kingdom

Any construction in the UK will require permission from the local council. The local planning authorities will each have their own specific regulations, so it is necessary to contact them before design and planning. The list of documents provided below will give you a head start, and they will be able to inform you if anything further is required.

United States

For most places in the U.S., construction requires a building permit. To obtain a building permit, first, contact the local public works department. They will be able to inform you of the zoning status and requirements of your zone. With this information in hand, you can tailor your design to the regulations required.

If your building site is outside the city's zoning code, then it may not require a building permit. If you seek to build without a permit, then deliberately selecting a site outside of the zoning code is one way to avoid a bit of red tape. However, it should be remembered that these sites will have less access to power, water, and telephone lines, and will thus present other challenges.

General List of Documents Required

Though each council or local authority will have its own specifications and regulations, here are a few things you can expect to need. Remember that regulations may influence aspects of design, so it still helps to approach your local authority before spending valuable time and energy finalizing your design.

- Structural engineering plans and approval
- Site plan
- Building regulation drawings (to scale)
- Before and after elevations
- Fully dimensioned working drawings

Designing Your Home

So, once you've considered your needs and done your homework concerning the local regulations, it's time to begin designing your home. Here is where you can get really creative. The simplest design would be a single container home, and the sky's the limit on how far you can go with it. Two-story? Three? The possibilities are endless, and they can be customized to fit your needs, whatever they might be.

It's most common to stick with single-story homes and to place containers next to one another until you reach your desired size. Connecting walls can be removed to increase the floor space of a room, and interior walls can be added to partition a container into multiple rooms. The basics are the bedroom, living room, kitchen, bathroom, and pantry, and all of these can be fit within a single 20-ft container if you're comfortable with a cozy living space.

A quick search online will offer many free software downloads that will help you to design your shipping container home. If you're uncomfortable doing this yourself, then you may want to factor the price of an architect into the budget. Usually, the quotes for such a small living space won't be unreasonable. However, give it a try and see how it works for you. You may be surprised how easy it is to design your home, and being in the driver's seat of the design process can be really fun.

Sample Plans

These are just a few of the designs we could show you to showcase the potential for customization in shipping container homes. Have a look and let inspiration strike!

Plan 1

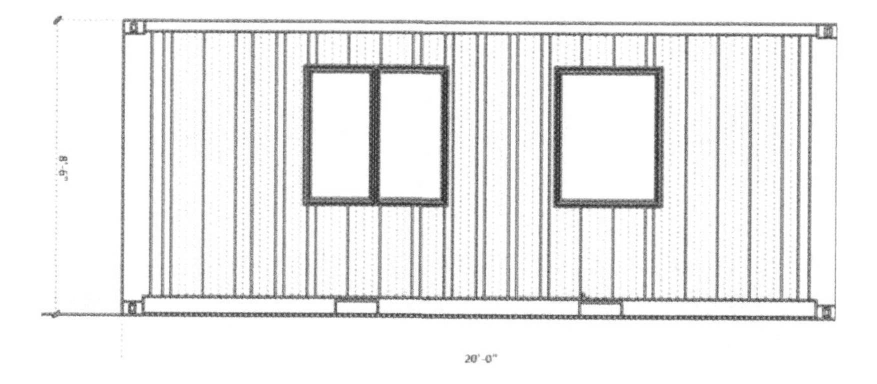

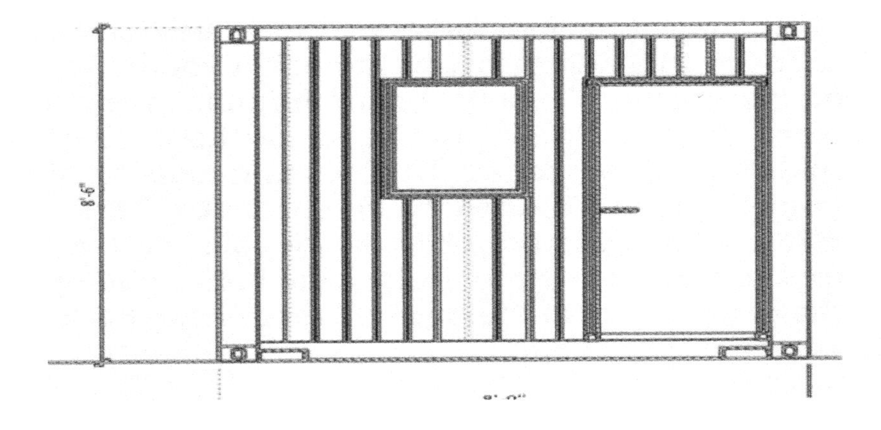

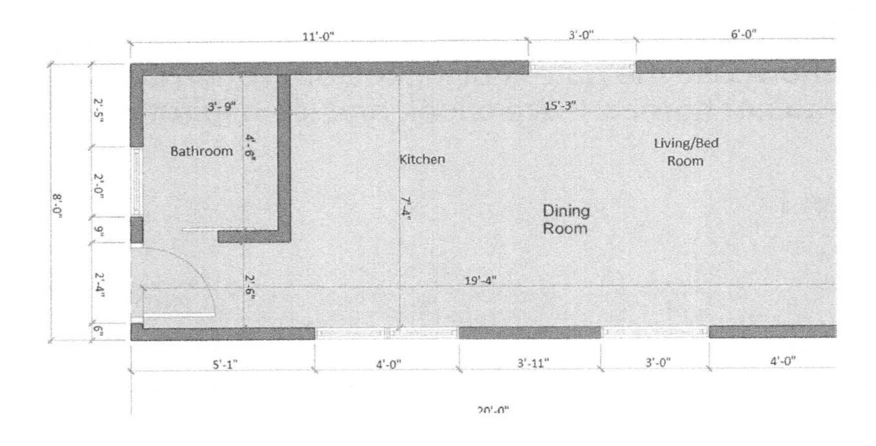

Plan 1 is ideal for a small, one-person dwelling. It offers 139 sq. ft. and is designed from a single 20-ft container and features a combination kitchen, dining room, living room, and bedroom—the ultimate in space efficiency and tiny living space.

Plan 2

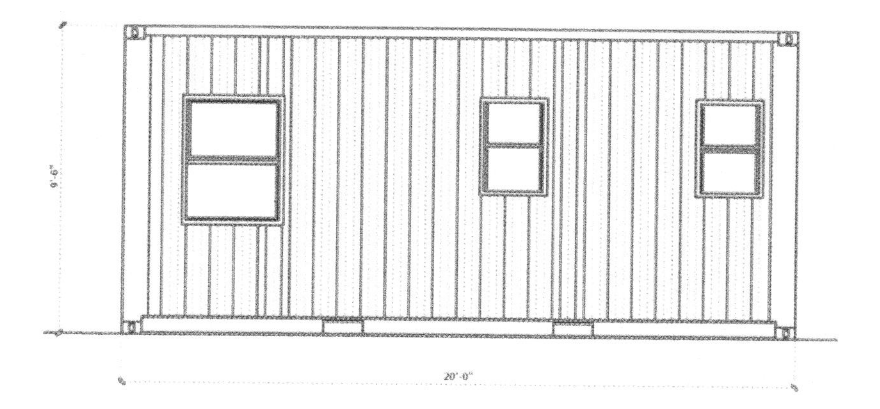

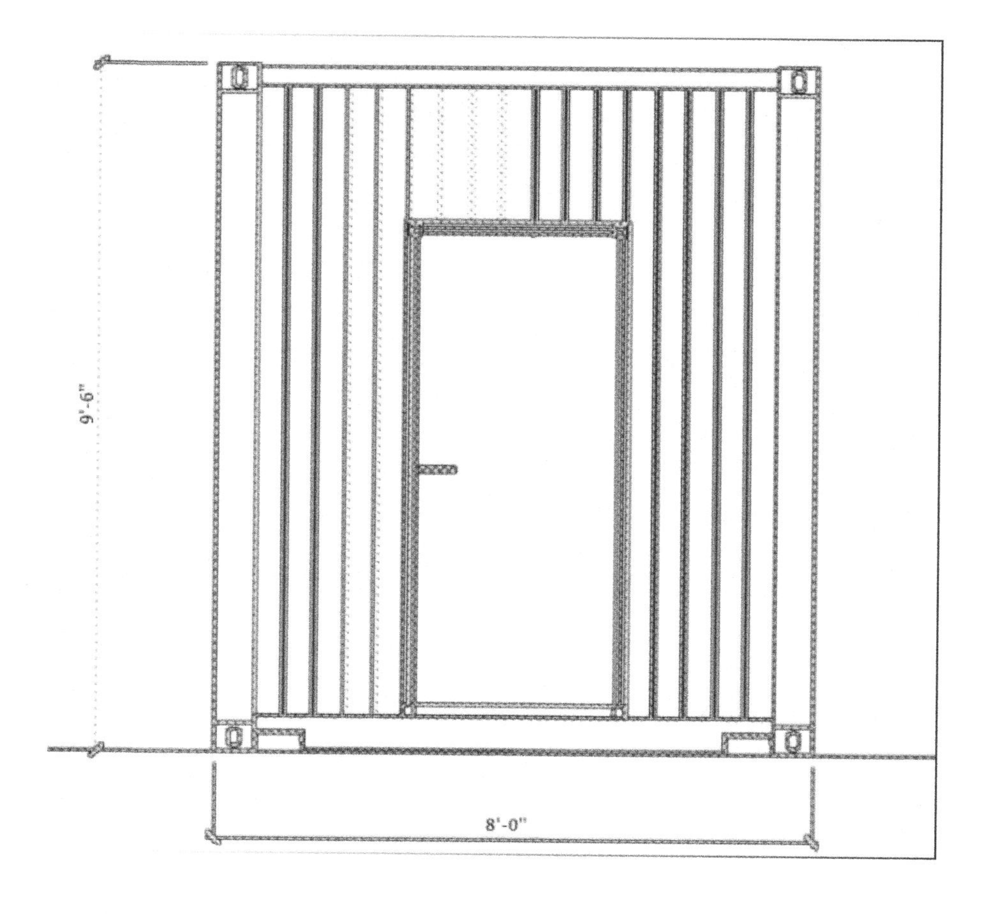

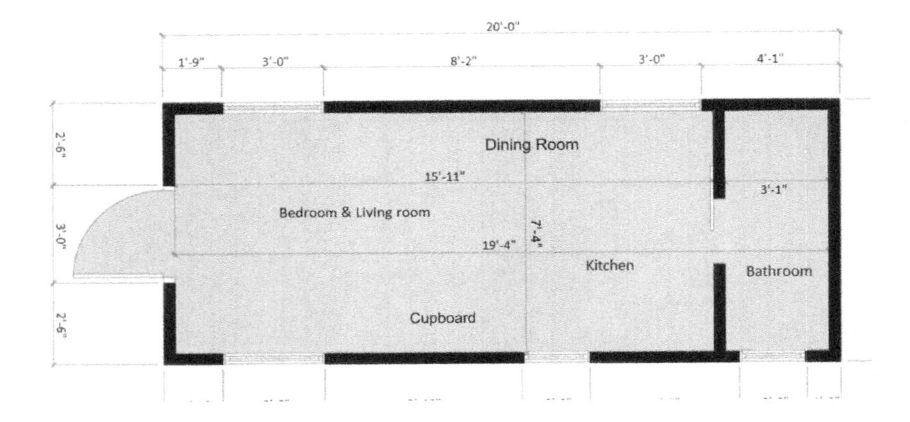

Plan 2 is another example of a single 20-ft container dwelling. It has been customized to offer a spacious bathroom and kitchen. The living room doubles as a bedroom and features two pull-out double beds, offering comfort whether sleeping or sitting. This economical design can comfortably house two people within 138 sq. ft.

Plan 3

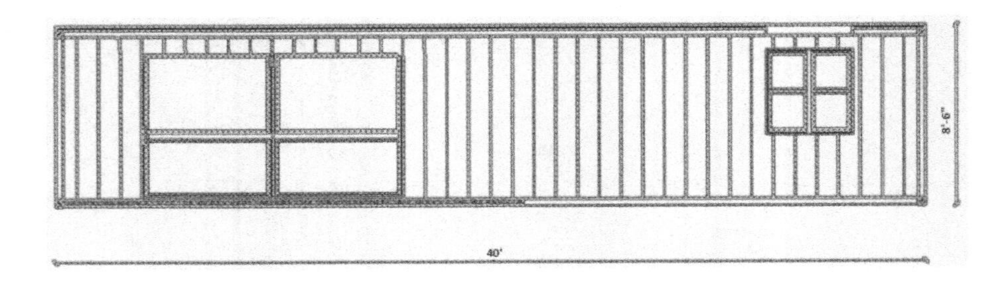

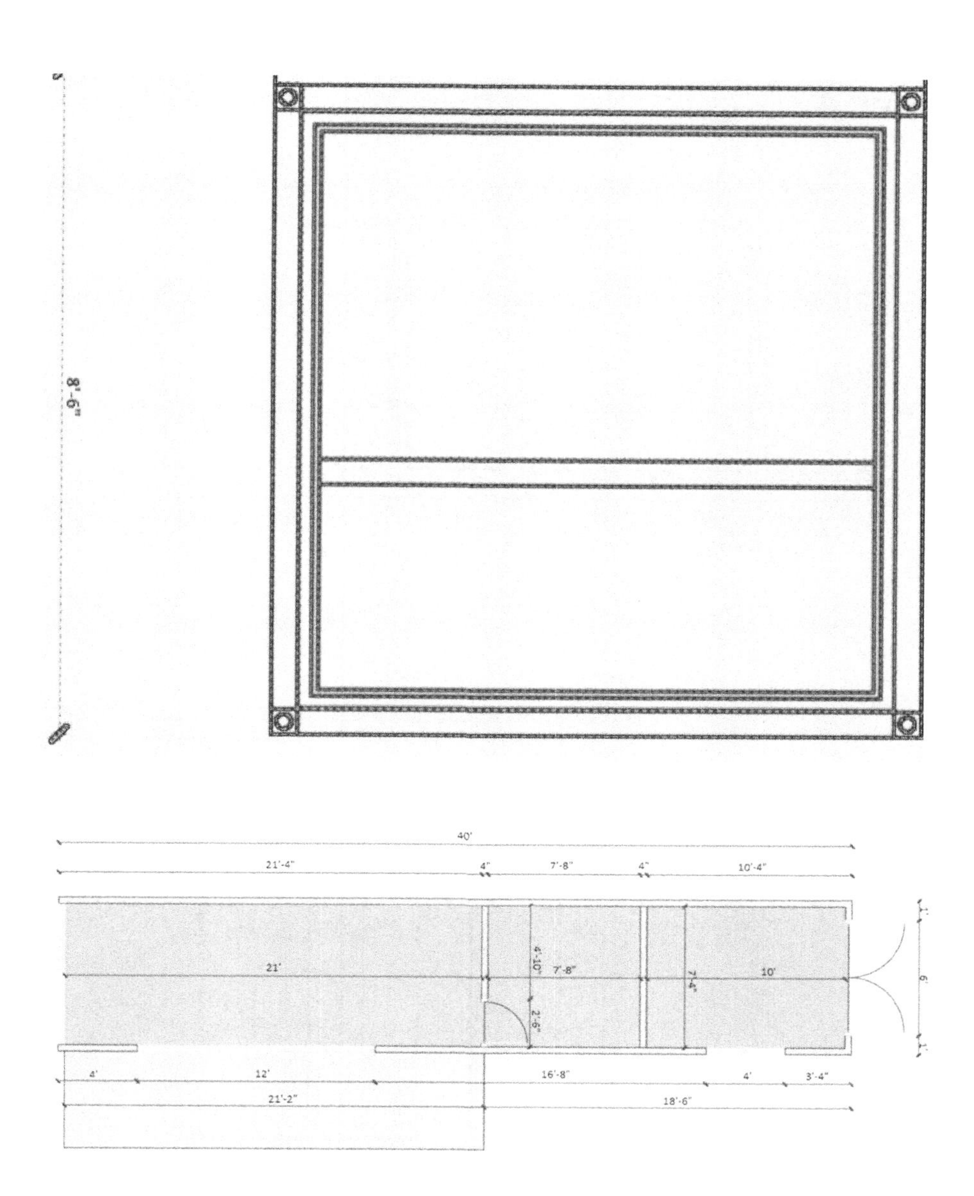

Plan 3 is designed from a single 40-ft. container and boasts a spacious open-plan living room which would be ideal for a sofa-bed. It features sliding glass doors which lead to an open deck. Also included is a second room ideal either for storage space or for a second bedroom.

Plan 4

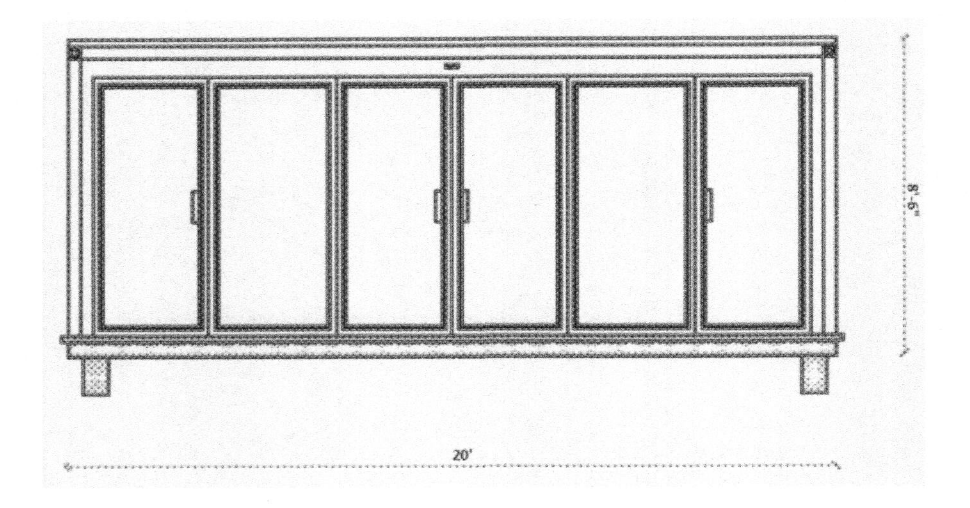

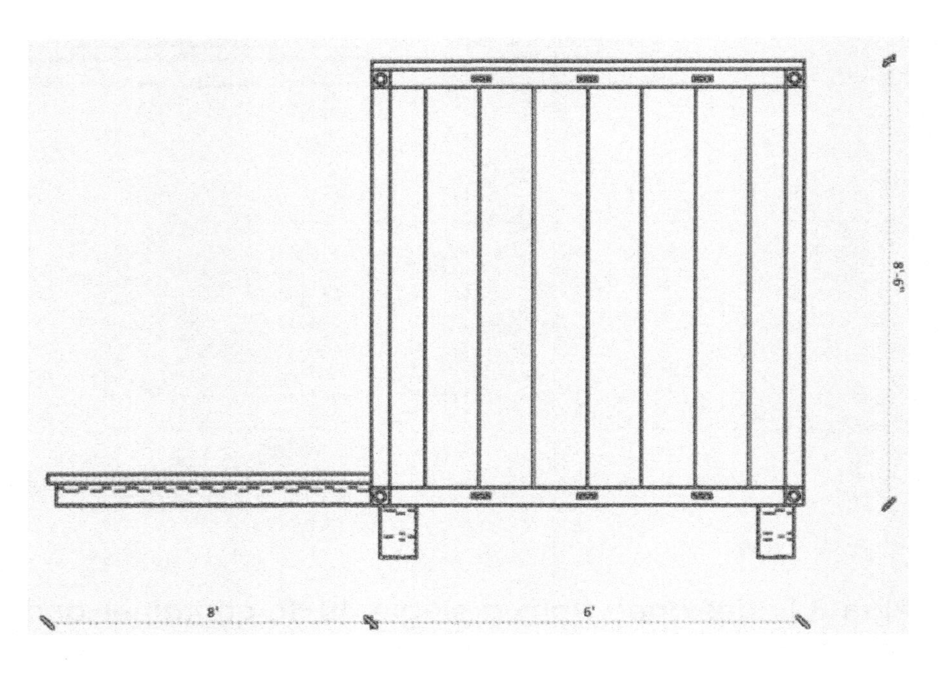

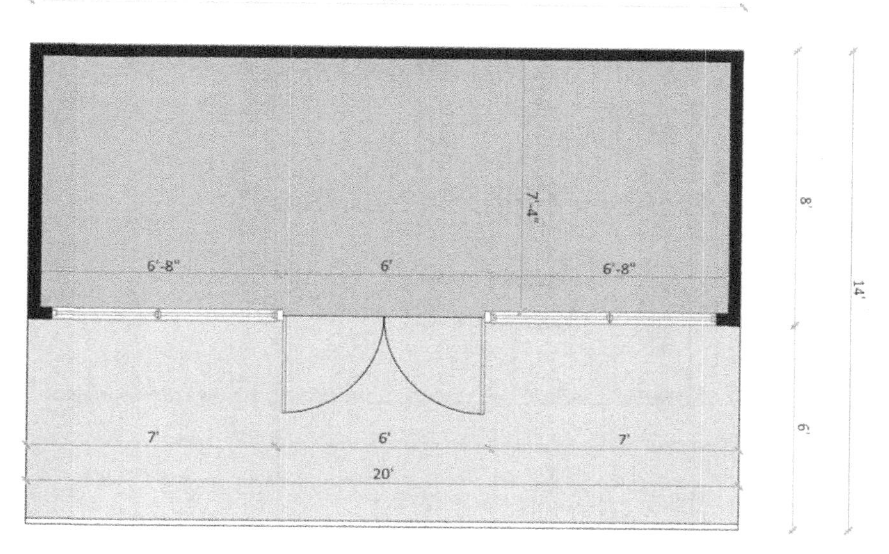

Plan 4 has been designed as a hunting lodge from a single 20-ft. container. It features a large open-plan living room with a small kitchen and bedroom off to the side. A full-length deck lines the front of the house, making it ideal for relaxing comfortably and taking in the night air. This offers a luxurious space for a single-person dwelling and can offer a comfortable space for two if a sofa bed is placed in the living room.

Plan 5

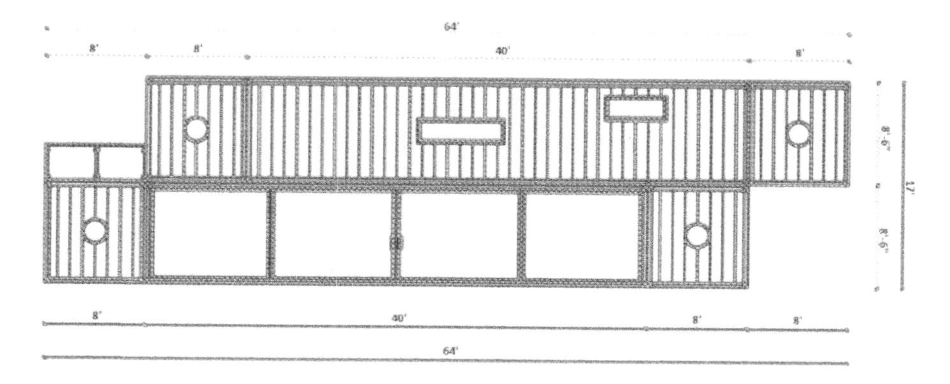

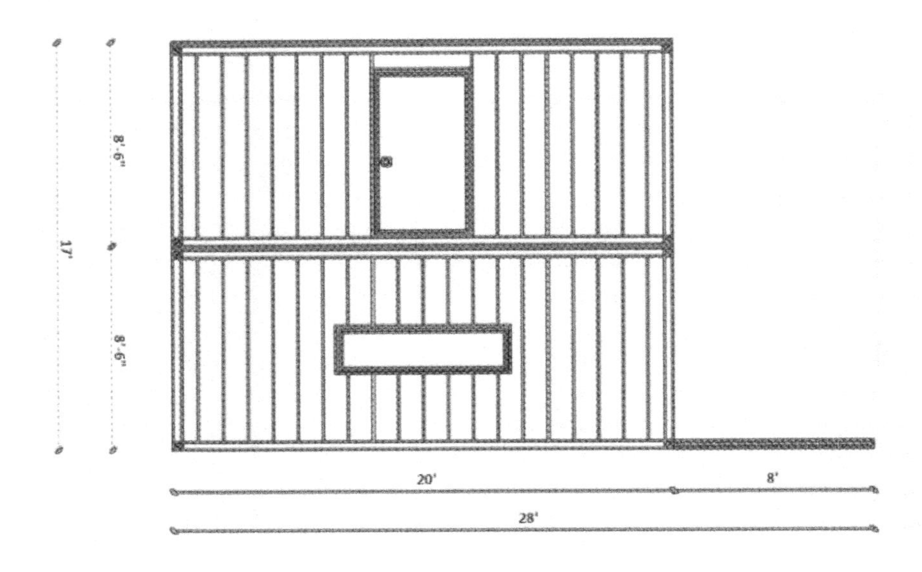

8'-6"
8'-6"
17'
20'
8'
28'

7'-4"
10'
20'
10'
7'-4"
1 2 3 4 5 6 7 8 9 10
40'
28'
20'

Upstairs:

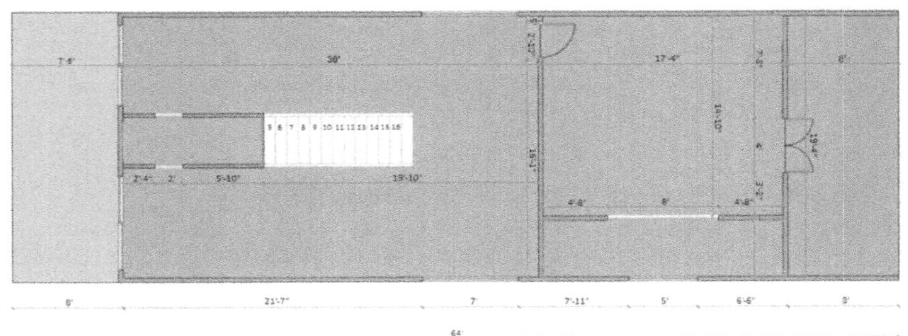

7'-4"
30'
17'-4"
5 6 7 8 9 10 11 12 13 14 15 16
2'-4"
2'
5'-10"
13'-10"
4'-8"
8'
4'-8"
8'
21'-7"
7'
7'-11"
5'
6'-6"
8'
64'

Plan 5 has been designed from four 40-ft. containers and four 20-ft. containers. It demonstrates the luxury possible with shipping container homes. This mansion features two floors, three bedrooms, three bathrooms, and an open-plan living room on both floors. Sliding glass doors on both the front and rear entrance, and a second-floor deck makes this design equal to any traditional home built with the height of decadence.

Plan 6

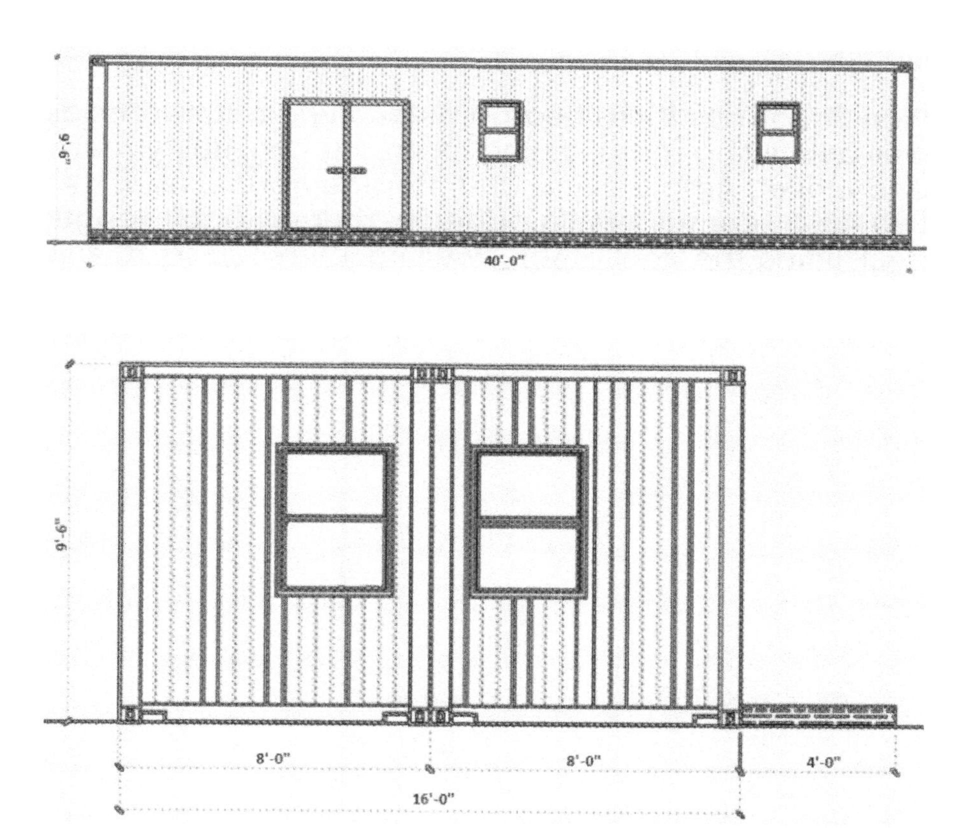

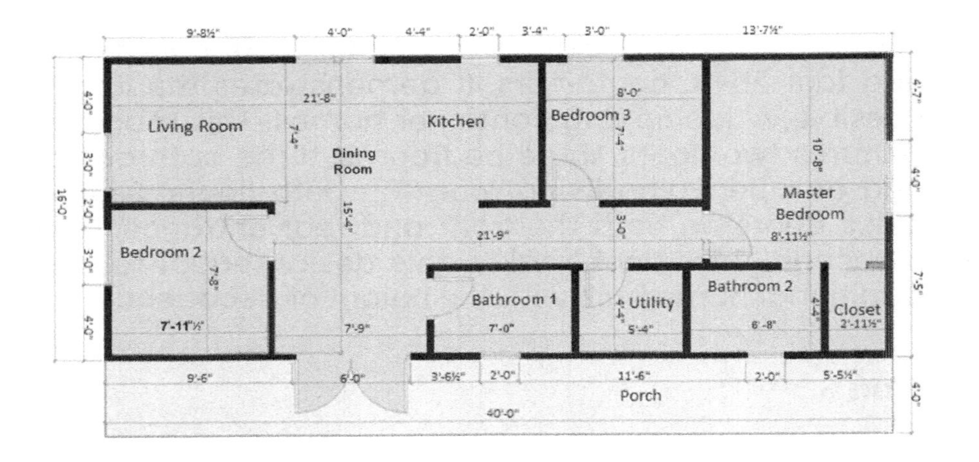

Plan 6 features three bedrooms and two bathrooms as well as a living room, dining room, kitchen, utility room, and closet. Ample living and storage space, all within two 40-ft. containers. In addition, there is a full-length deck lining the front of the dwelling. With 606 sq. ft., this dwelling can comfortably house three adults or a family of four.

Plan 7

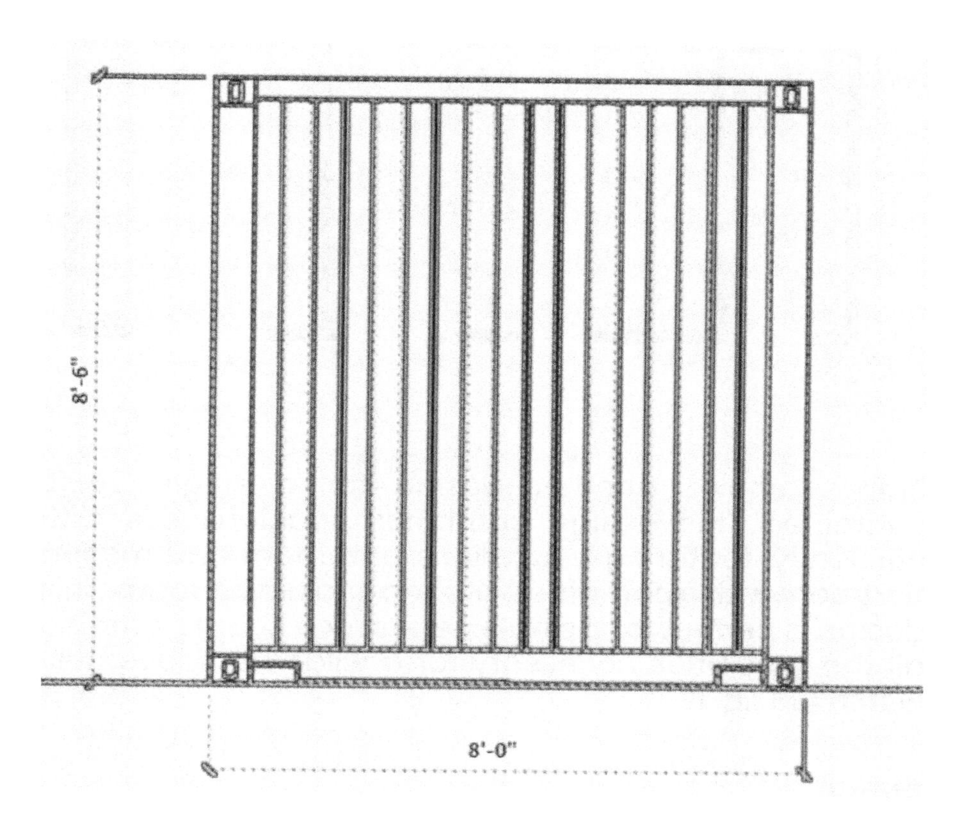

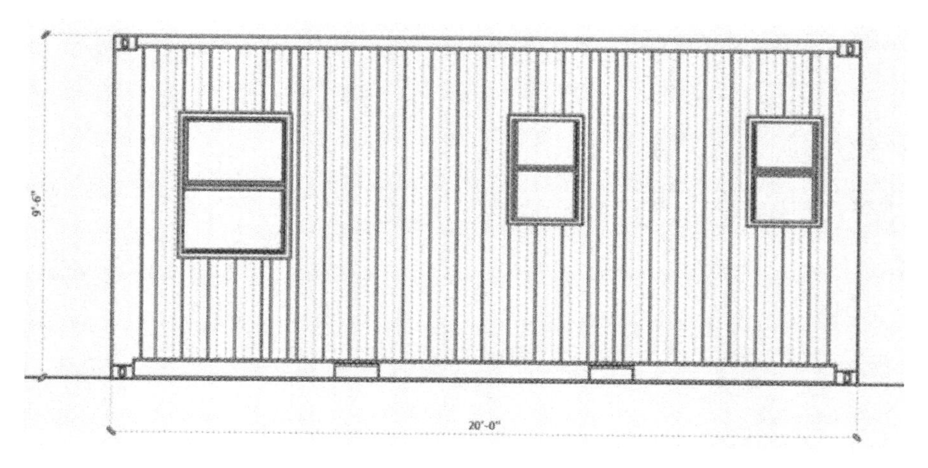

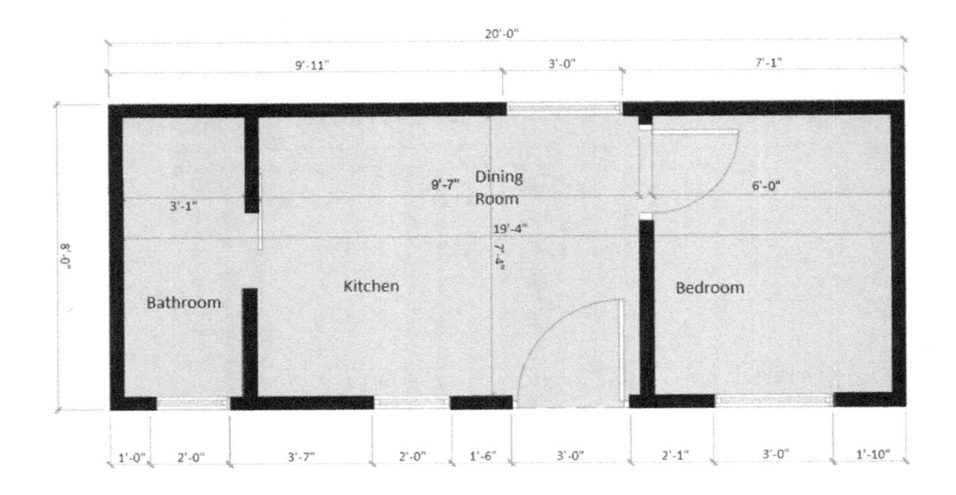

Plan 7 shows another design for a single 20-ft. container. This design combines spaciousness with efficiency, featuring a luxurious bathroom and master bedroom with a combined kitchen and dining room. This design is perfect for a single person or a couple, offering all the amenities necessary for a wilderness love nest within 135 sq. ft.

Plan 8

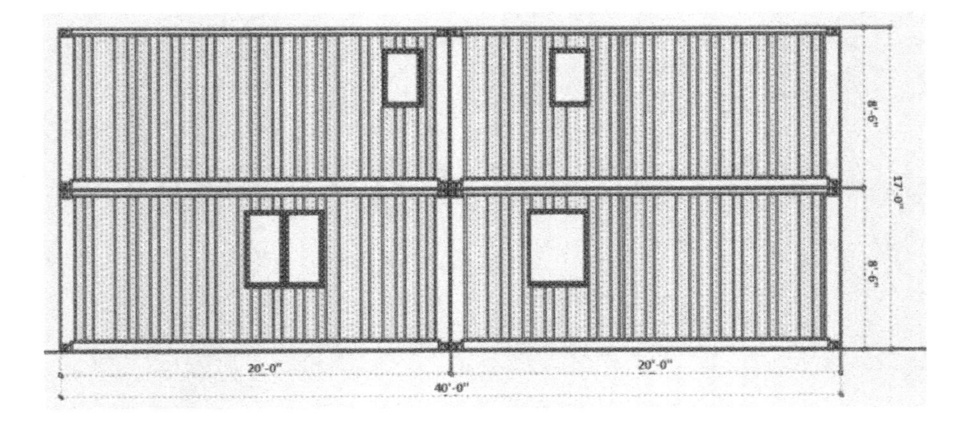

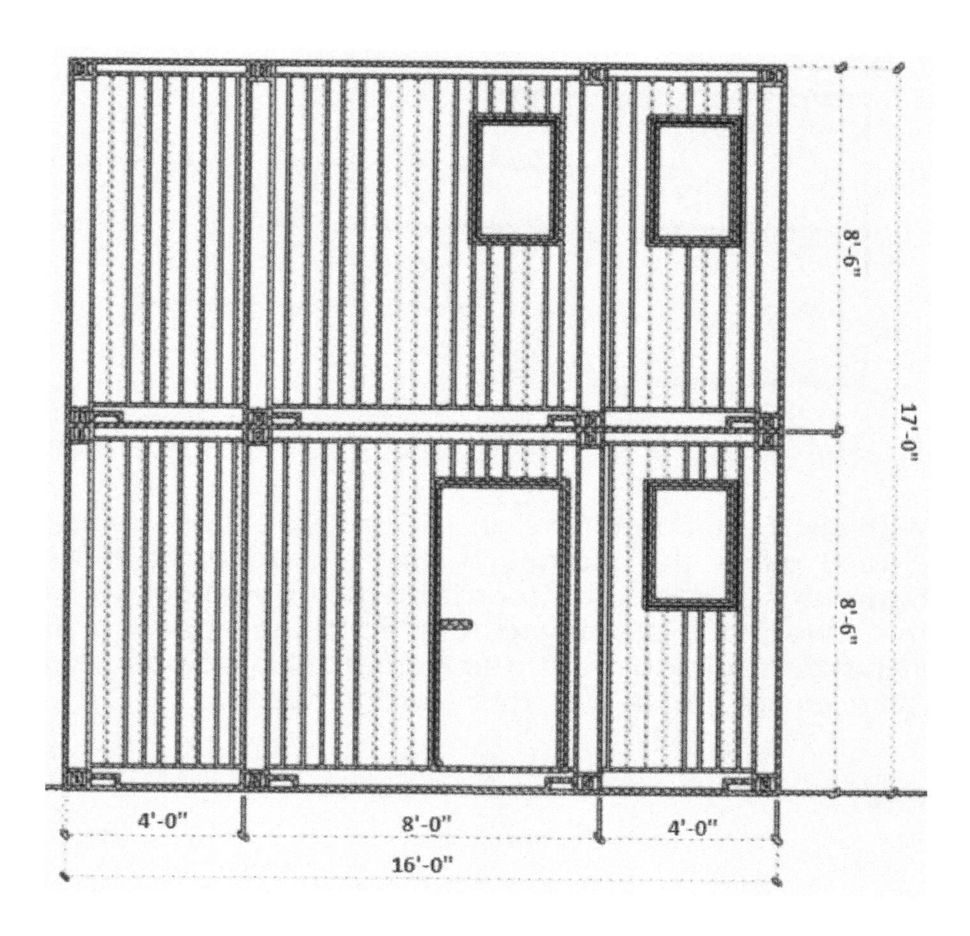
8'-6"
17'-0"
8'-6"
4'-0"
8'-0"
4'-0"
16'-0"

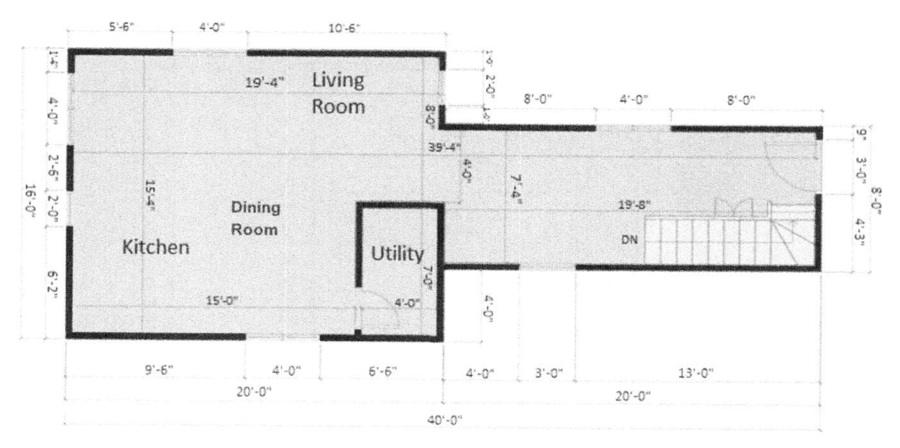
5'-6"
4'-0"
10'-6"
Living Room
19'-4"
8'-0"
4'-0"
8'-0"
39'-4"
7'-4"
19'-8"
DN
Dining Room
Kitchen
Utility
15'-4"
15'-0"
16'-0"
9'-6"
4'-0"
6'-6"
4'-0"
3'-0"
13'-0"
20'-0"
20'-0"
40'-0"

Upstairs:

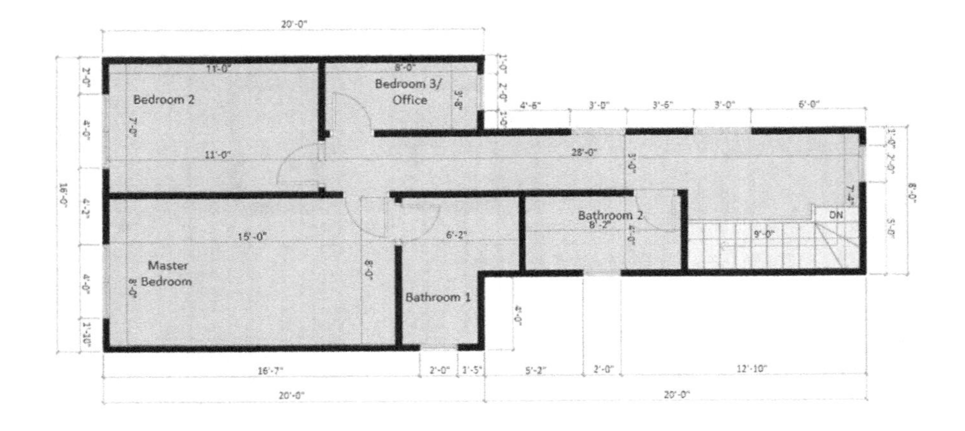

With six 20-ft. containers and 861 sq. ft. of floor space, Plan 8 offers the utmost of luxury. It features three bedrooms and two bathrooms on the upper floor, while the lower floor is devoted to a combined open-plan living room, dining room, and kitchen. Also included is a utility closet that serves your storage needs.

Plan 9

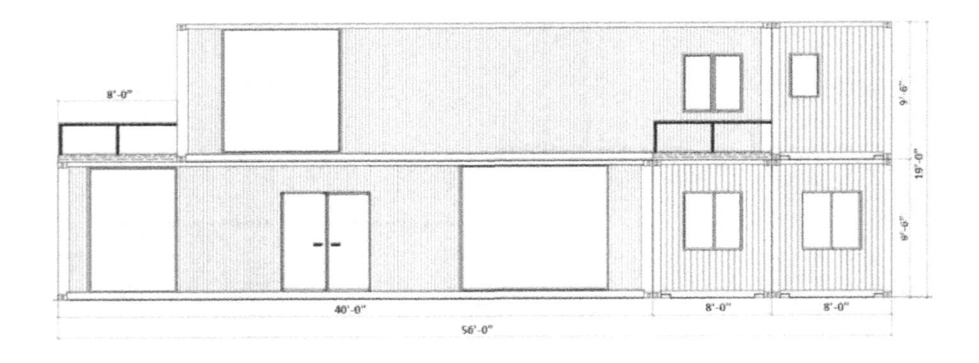

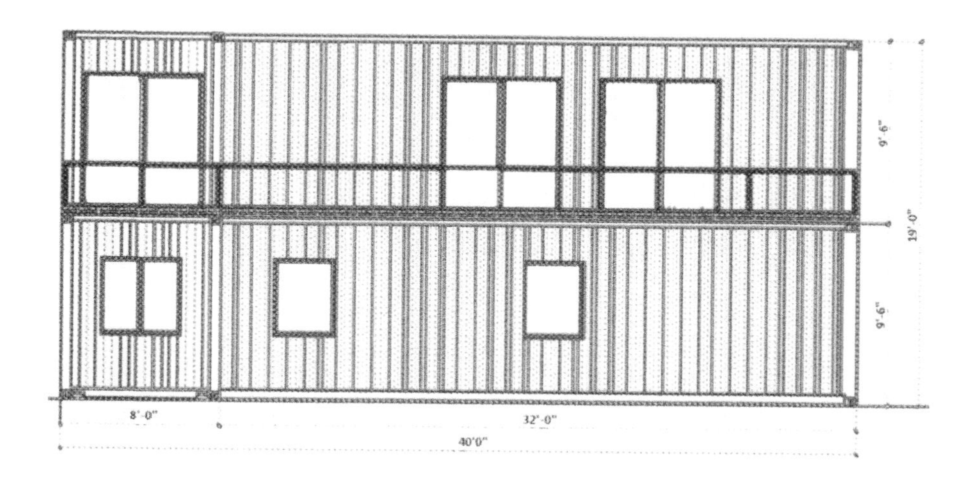
9'-6"
19'-0"
9'-6"
8'-0"
32'-0"
40'0"

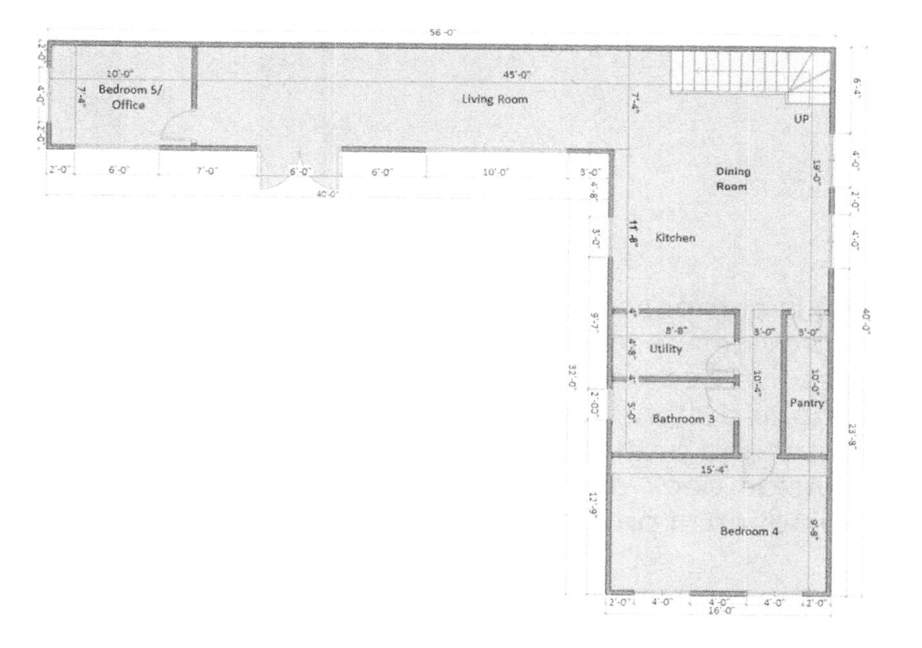
56'-0"
Bedroom 5/
Office
Living Room
45'-0"
Dining
Room
Kitchen
Utility
Bathroom 3
Pantry
Bedroom 4
UP
40'-0"
32'-0"
16'-0"
15'-4"

Upstairs:

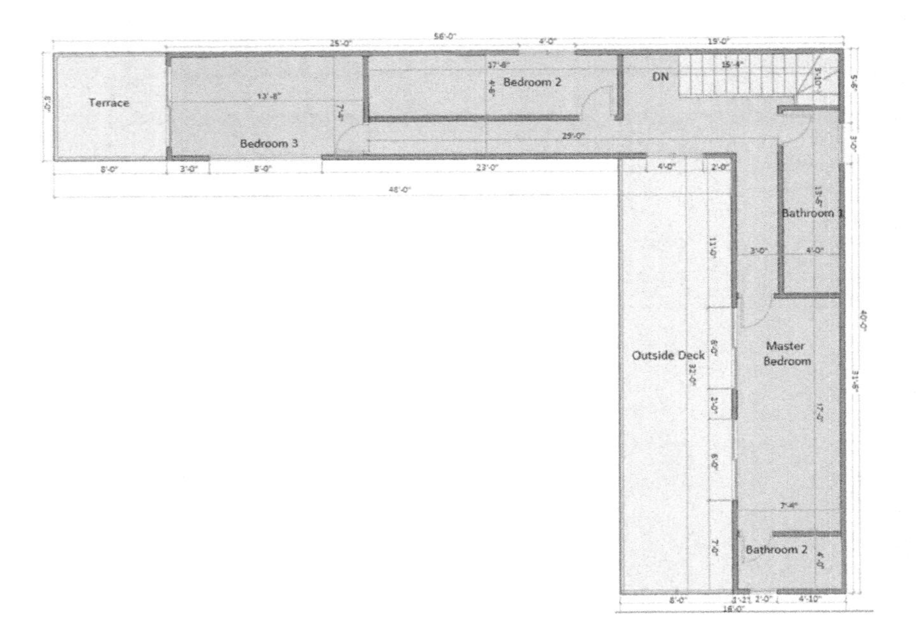

Plan 9 is composed of five 40-ft. high cube containers and provides a total of 1718 sq. ft. of floor space. It houses five bedrooms and three bathrooms as well as a combined kitchen and dining room. Also are included a pantry and a utility closet. The second floor features an outside deck and terrace, which are perfect for a combination of privacy and fresh air.

Plan 10

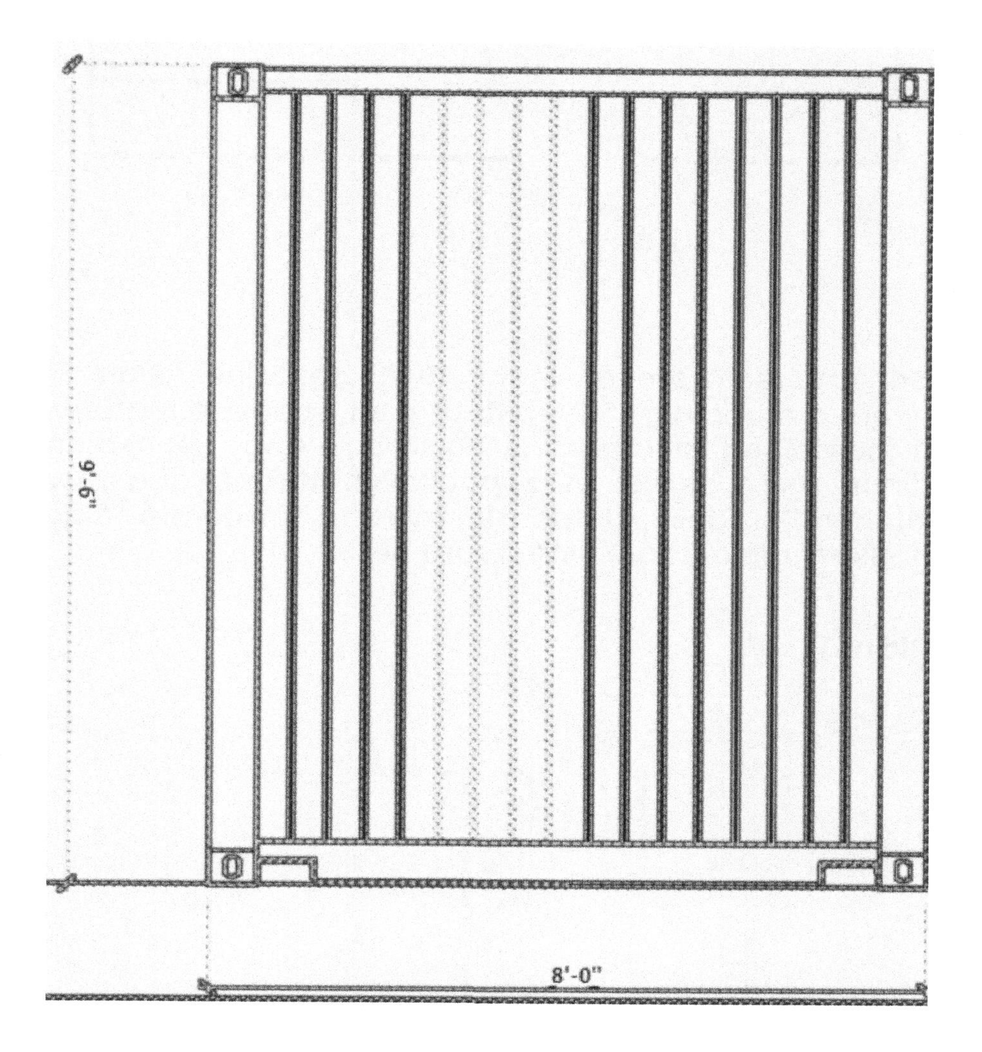

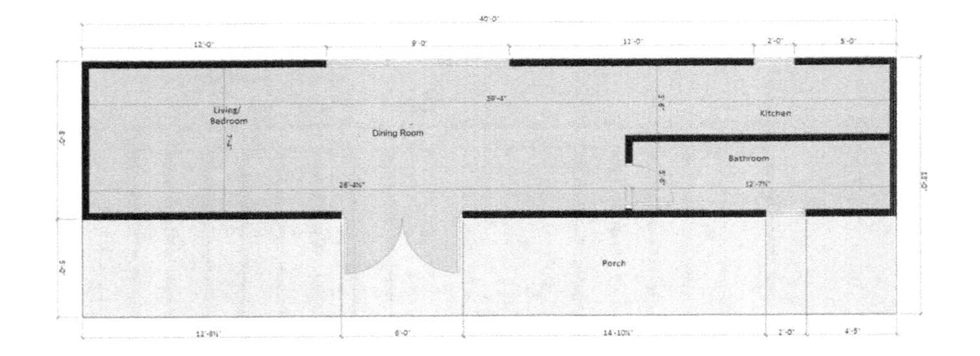

Constructed from a single 40-ft. container, Plan 10 offers a stunning 483 sq. ft. of floor space. It features a combined bedroom/living room and open-plan dining room, as well as a spacious bathroom and cozy kitchen. This design is ideal for two and can comfortably house three or more with a sofa bed.

Plan 11

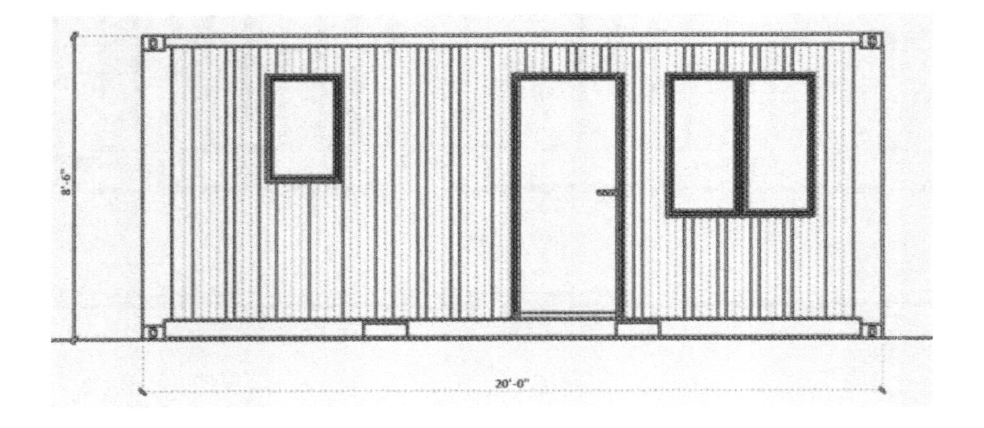

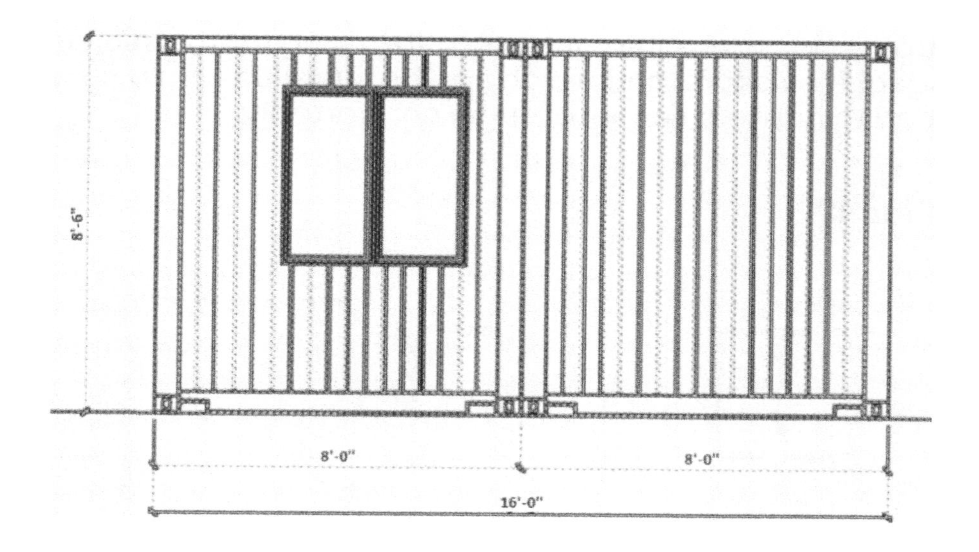

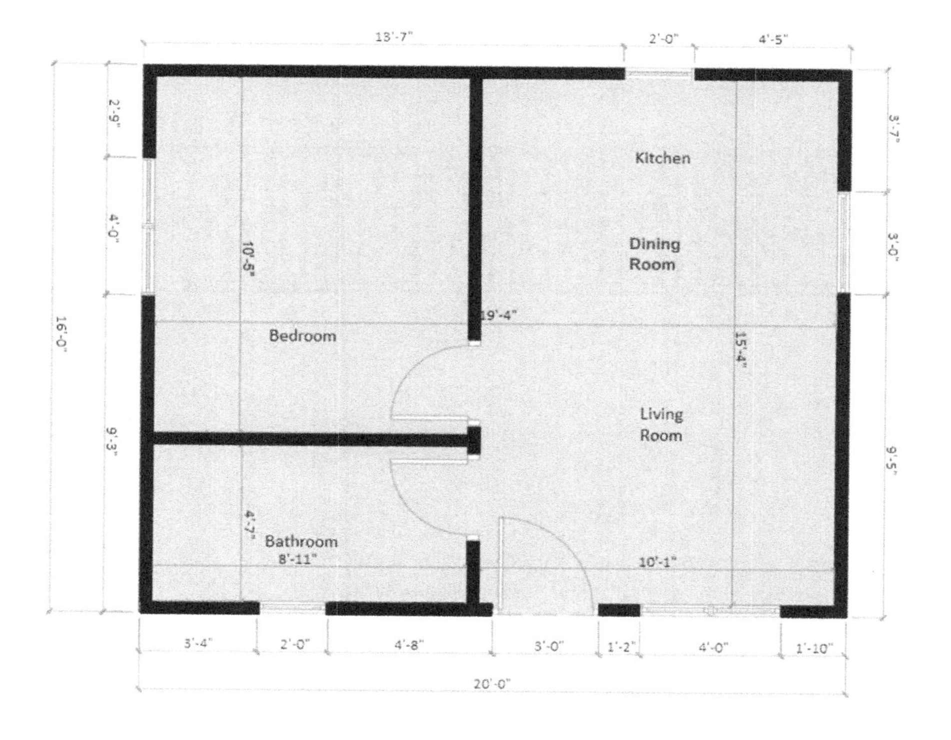

Made from only two 20-ft. containers, Plan 11 offers 289 sq. ft. of floor space. It features a master bedroom, luxurious bathroom, and combined kitchen and dining room. The front door opens into an open-plan living room which is ideal for visitors, and which can be supplied with a sofa bed to house a second person comfortably.

Plan 12

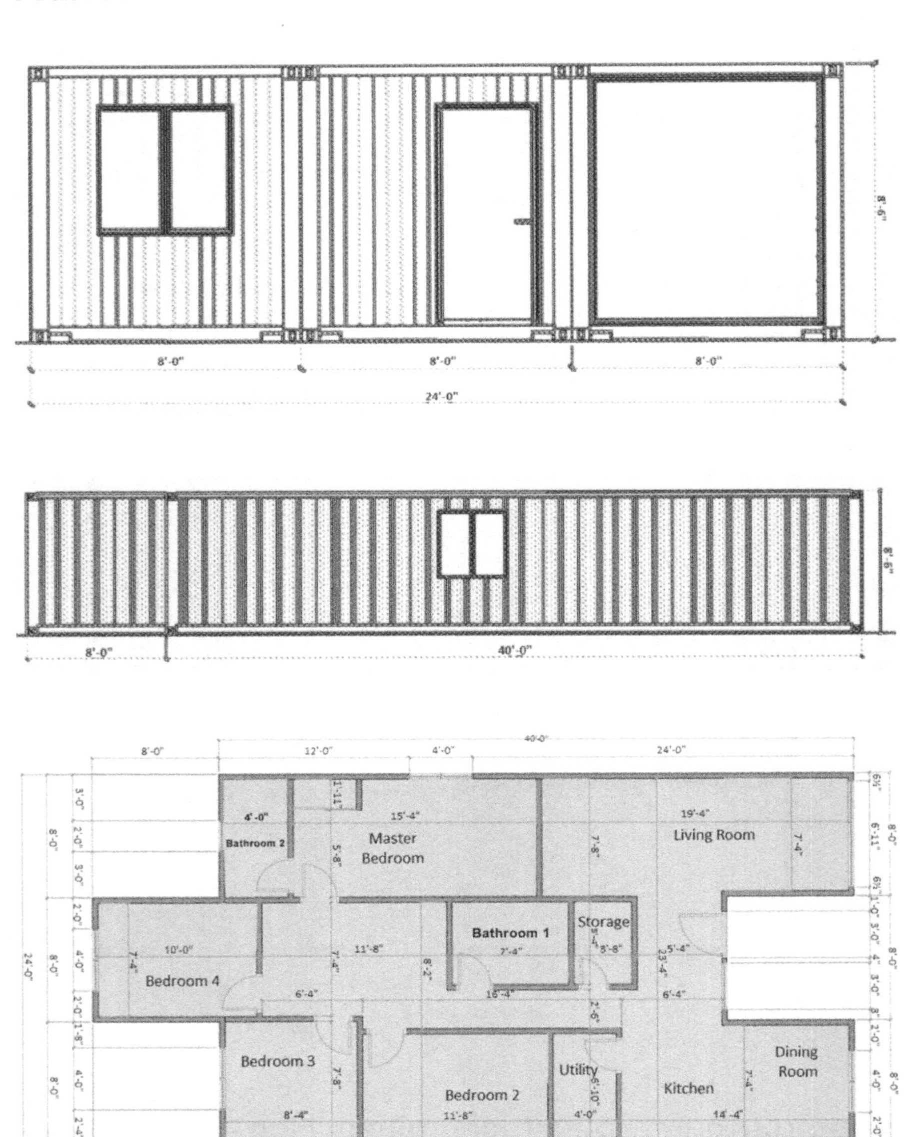

Constructed from merely three 40-ft. containers, Plan 12 offers a spacious 899 sq. ft. of floor space. It features a master bedroom as well as three additional bedrooms and two bathrooms. It also has an efficiently combined dining room and kitchen as well as a separate living room with ample space for entertaining visitors.

Planning Checklist

- Plan out the needs of your home.
- Set your budget.
- Meet with the local planning authorities to find the specifications and required paperwork for your area.
- Design your home.
- Obtain planning permission, if necessary.

CHAPTER 6
Insulation

One of the main disadvantages of a container house is its lack of insulation.

Let's face it, it's a metal box. If you leave it under the sun for too long, it's going to bake everything inside it. When the snow comes, it becomes an icebox. Living in a house with such temperature extremes will prove to be very challenging, even if you're very passionate about container homes.

So, what are you going to do about it?

Well, first you have to assess your situation. Do you live in a hot or cold area? There are different ways to address temperature issues in your container house (or any house for that matter).

If you live in a temperate area, you're not going to insulate it heavily. Instead, the best solution is to find a way to decrease the exposure to the sun and modify your building to increase the airflow.

Your typical container house would have more spaces cut out of it to give room to bigger doors and windows.

On the other hand, if you live in a cold area, it would be in your best interest to retain as much heat in your house as possible. One feature that you could include in your house is a wood stove or an electric heater. It would also be more practical to have smaller windows and doors so if and when you open them, only a minimal amount of heat would escape from these portals. Another option to go for is to properly insulate your entire container house.

Insulating Container Houses for Tropical Countries/Areas

It would be in your best interest to increase the airflow into your container house as well as decrease the amount of exposure to the sun it gets.

Here are some solutions that you can use to address heat issues:

Increase Size of Openings to Increase Airflow

A basic shipping container has one large opening composed of two swinging doors at one end. This is where the ingress and egress of materials often happen.

For your shipping container to pass several building code requirements, it must have another exit point.

Here's something you should remember: Cutting into the sides of your shipping container will weaken its base structural strength. But that can be remedied by reinforcing the cut areas with steel beams and posts. This means you can cut any size of doors or windows on your shipping container as long as you make sure that you immediately reinforce the openings to prevent the entire structure from collapsing.

Decrease the Structure's Exposure to the Sun

Doctors often advise their patients not to stay out in the sun for too long to reduce the risk of heatstroke. The solution to this would be to stay indoors or get under the shade of a tree or a building as soon as you can.

This is going to be a problem for a container house since, although it is portable, it's not easy (or practical) to keep moving it.

The more your container house bakes under the sun, the more it's going to affect the people and things inside it.

So, how can you solve this issue?

Over the years, there have been great solutions provided by people on how to keep your container house cool.

- Trees

These are the best solutions to keeping your house cool. It's also the greenest solution of all because it's all-natural.

Situating your house under a tree takes advantage of the natural shade provided by the branches and leaves and it is very aesthetically pleasing as well.

If you don't have trees around, you can transplant some and strategically situate them in your property to take care of the heat issue.

This is the best answer to keeping your house cool by reducing the exposure to the sun.

- Install a raised roof

Another ingenious solution to keeping your sheltered from too much sun is installing a raised roof. This will catch the majority of the sun's rays and trap most of the heat in between the raised roof and the container house. By keeping it open, the air can carry off the excess heat away.

- Install a canopy

Some container house owners have found that installing a large canopy over their houses can greatly lessen sun exposure. This solves most of the cooling issues, and it is a cost-effective solution for your container house.

Insulating Container Houses for Cold Countries/ Areas

For all its strengths, a container home does have a few caveats. One of these is poor heat insulation. Once it gets cold, there's no way to stop from freezing your toes off unless you properly insulate your house.

Insulating Material for Container Houses

Foam Insulation

The best insulation for container houses would be sprayed foam insulation. If you've ever been inside a shipping container, you'll notice that the corrugation that you often see outside extends inside as well. This corrugated wall is one reason why a shipping container's sides are so strong.

You can use conventional foam and cut it to the proper shape to insulate your container, but that will be a very taxing and time-consuming endeavor (it's still a good idea, though).

Using foam spray is going to cut that process down to more than half the time it takes to insulate your house. All you need to do is put the wall cladding up and then leave a small space for the hose to fit. Spray the foam in and wait for it to fill in the entire space between metal and wood.

Since it comes in a semi-liquid/ foam form, it has a better chance to conform to the shape of the metal. This allows for lesser gaps between the metal and the wood, thereby trapping heat and keeping it in much more effectively.

This process is the best solution, but it's going to cost you a pretty penny.

Hay

Stacking bales of hay for insulation is also a good option. Most conventional houses also use this method.

You can also choose to stack hay against your outer walls and either leave it exposed or cover it with a layer of wood.

Leaving it exposed isn't that great of an idea since hay can trap the moisture in. just remember that the hay is going to be flush against your metal exterior and too much exposure of metal to moisture can lead to rusting.

Hay is inexpensive and can be replaced over time without any major issues and it softens the look of a container house.

Covering the hay with a façade is another option you

can take if you're going to use this material. Your hay is still going to be on the outside of your container house but it's going to be covered with another wall.

Burying Your Container House

If you like this idea, you're probably into prepping as well because it's essentially going to turn your house into a bunker.

Packed earth is actually a very good insulating material. But this process is going to cost you a lot because you will have to excavate the area where you are going to bury an entire shipping container. The bigger your shipping container is, the bigger the hole you'll need. And the bigger the hole it's going to burn into your pocket!

Still, an excellent idea but could lead you to the poor house.

There are options, though, if you want to bury your container house. You can choose to bury only a portion of it or base it on the landscape your house is going to be in.

CHAPTER 7
Heating and Cooling of the House

One of the main disadvantages of a container house is its lack of insulation.

Let's face it, it's a metal box. If you leave it under the sun for too long, it's going to bake everything inside it. When the snow comes, it becomes an icebox. Living in a house with such temperature extremes is going to prove to be very challenging, even if you're very passionate about container homes.

So, what are you going to do about it?

Well, first, you have to assess your situation. Do you live in a hot or cold area? There are different ways to address temperature issues in your container house (or any house for that matter).

If you live in a temperate area, you're not going to insulate it heavily. Instead, the best solution is to find a way to decrease the exposure to the sun and to modify your building to increase the airflow.

Your typical container house would have more spaces cut out of it to give room to bigger doors and windows.

On the other hand, if you live in a cold area, it would be in your best interest to retain as much heat in your house as possible. One feature that you could include in your house is a wood stove or an electric heater. It would also be more practical to have smaller windows and doors so, if and when you open them, only a minimal amount of heat would escape from these portals. Another option to go for is to properly insulate your entire container house.

How to Cool the House in Summer?

It would be in your best interest to increase the airflow into your container house as well as decrease the amount of exposure to the sun it gets.

Here are some solutions that you can use to address heat issues:

- Increase the size of openings to increase airflow

A basic shipping container has one large opening composed of two swinging doors at one end. This is where the ingress and egress of materials often happen.

For your shipping container to pass several building code requirements, it must have another exit point.

Here's something you should remember: Cutting into the sides of your shipping container will weaken its base structural strength. But that can be remedied by reinforcing the cut areas with steel beams and posts. This means you can cut any size of doors or windows on your shipping container as long as you make sure that you immediately reinforce the openings to prevent the entire structure from collapsing.

- Decrease the structure's exposure to the sun.

Doctors often advise their patients not to stay out in the sun for too long to reduce the risk of heat stroke. The solution to this would be to stay indoors or get under the shade of a tree or a building as soon as you can.

This is going to be a problem for a container house since, although it is portable, it's not easy (or practical) to keep moving it.

The more your container house bakes under the sun, the more it's going to affect the people and things inside it.

So, how can you solve this issue?

Over the years, there have been great solutions provided by people on how to keep your container house cool.

- **Trees** – these are the best solutions to keeping your house cool. It's also the greenest solution of all because it's an all-natural.

Situating your house under a tree takes advantage of the natural shade provided by the branches and leaves, and it is very aesthetically pleasing as well.

If you don't have trees around, you can transplant some and strategically situate them in your property to take care of the heat issue.

This is the best answer to keeping your house cool by reducing the exposure to the sun.

- **Install a raised roof**

Another ingenious solution to keeping your sheltered from too much sun is installing a raised roof. This will catch the majority of the sun's rays and trap most of the heat in between the raised roof and the container house. By keeping it open, the air can carry off the excess heat away.

- **Install a canopy**

Some container house owners have found that installing a large canopy over their houses can actually lessen the sun exposure greatly. This solves most of the cooling issues, and it is a cost-effective solution for your container house.

How to Heat the House in Winter?

For all its strengths, a container home does have a few caveats. One of these is poor heat insulation. Once it gets cold, there's no way to stop from freezing your toes off unless you properly insulate your house.

Insulating material for container houses

- **Foam Insulation**

The best insulation for container houses would be sprayed foam insulation. If you've ever been inside a shipping container, you'll notice that the corrugation that you often see outside extends inside as well. This corrugated wall is one reason why a shipping container's sides are so strong.

You can use conventional foam and cut it to the proper shape to insulate your container but that is going to be a very taxing and time-consuming endeavor (it's still a good idea though).

Using foam spray is going to cut that process down to

more than half the time it takes to insulate your house. All you need to do is put the wall cladding up and then leave a small space for the hose to fit. Spray the foam in and wait for it to fill in the entire space between metal and wood.

Since it comes in a semi-liquid/ foam form, it has a better chance to conform to the shape of the metal. This allows for lesser gaps between the metal and the wood, thereby trapping heat and keeping it in much more effectively.

This process is the best solution, but it's going to cost you a pretty penny.

- **Hay**

Stacking bales of hay for insulation is also a good option. Most conventional houses also use this method.

You can also choose to stack hay against your outer walls and either leave it exposed or cover it with a layer of wood.

Leaving it exposed isn't that great of an idea since hay can trap the moisture in. just remember that the hay is going to be flush against your metal exterior and too much exposure of metal to moisture can lead to rusting.

Hay is inexpensive and can be replaced over time without any major issues and it softens the look of a container house.

Covering the hay with a façade is another option you can take if you're going to use this material. Your hay is still going to be on the outside of your container house but it's going to be covered with another wall.

- **Burying your container house**

If you like this idea, you're probably into prepping as well because it's essentially going to turn your house into a bunker.

Packed earth is actually a very good insulating material. But this process is going to cost you a lot because you will have to excavate the area where you are going to bury an entire shipping container. The bigger your shipping container is, the bigger the hole you'll need. And the bigger the hole it's going to burn into your pocket!

Still, an excellent idea but could lead you to the poor house.

There are options though if you want to bury your container house. You can choose to bury only a portion of it or base it on the landscape your house is going to be in.

Containers are steel boxes; you need insulation to make them livable.

Why Do You Need Insulation in The First Place?

- **Heat Control:** Since the containers are steel structures, they absorb and transmit heat and cold. The container cannot be controlled if no insulation is used.

- **Humidity:** The moist of the interior air condenses against the cold steel. This will lead to condensation and after some time mold.

- **Temperature:** Steel is a good conductor of heat and cold. Insulation is needed to make the containers appropriate for you to live comfortably in without outrageous energy bills.

You might be wondering what kind of insulation is ideal for your container home. The type of insulation you should use depends on various things such as:

- Cost involved
- Your local climate
- Environment friendliness

You not only need to insulate the inside of your container, but painting the outside of your container can have a big difference.

If you live in a hot climate with the sun beating down on your house, you can consider ceramic insulation sprayed or painted on the outside of your container to reflect the heat. Ceramic paint has a good R-value.

Ceramic Paint

If you would like to go cheaper, you can use white paint. I prefer using white paint over ceramic insulation because of the cost involved.

Don't make the mistake of painting your shipping container black if you live in a hot climate. You will basically create an oven.

Here are some insulation types you can consider for your container:

Fiberglass Insulation (Glass Wool)

Fiberglass insulation is one of the common insulation methods in traditional homes. Fiberglass insulation is also widely used in container homes. The fiberglass is generally 3.5 inches thick and provides good protection against outside heat or cold and maintains indoor temperature. The fiberglass covers the ribbed sides of the walls as well as the electric wiring

If you don't wear protection, it will irritate your skin, eyes, and respiratory system. You need to know what you are doing If you are using fiberglass.

This insulation will come in rolls, and depending on the type, the fibers might be covered with paper (like packaging). Don't remove these, they will protect you from the fibers.

They can be hard to install if you are running quite a lot of electrical wires and plumbing through the framing. You will have to install a moisture barrier if you use this insulation.

You will need to install a vapor barrier (more in this later). It has an R-value of 3.5 per inch.

Styrofoam Insulation

It's also called blue board or close-cell extruded polystyrene foam (XPS). It's different from the classic white foam you see in the packaging. This is because the white foam is expanded while the blue Styrofoam is extruded. Its R-value is 5 per inch.

If you want to install Styrofoam into your framing, you need to finish the corners off with a spray foam can to seal the edges.

If you decide to go with Styrofoam, be prepared to cut a lot of boards because the boards won't fit the frames most of the time.

As with fiberglass insulation, it's quite hard and a lot of work to install Styrofoam along your walls. That's why I'm going to talk more about spray foam next. You will have to install a moisture barrier if you use this insulation.

Spray Foam

Spray foam also provides good insulation to container homes. The spray foam completely hides the ribs, plumbing, and electrical wiring. Unlike fiberglass or polystyrene panels, spray foam does not leave any gaps between the wall and insulation. Therefore, there is no chance of condensation or moisture developing.

The open cell variant provides an R-3,5 value

The closed-cell variant provides an R-6 value per inch. It's more expensive than the open-cell but has a great advantage besides it having a better R-value than the open cell.

Let's say you are using open spray foam and it's very cold outside (freezing). You are sitting cozy in your seat at room temperature. The air inside has more vapor in it than the cold air outside. Because the cells of the spray foam are open and not closed the warm air touches the cold surface and starts to condensation. Leading to rust and rotting.

If you would have used closed foam spray the air from the inside can't reach the outer wall because it's air tight. So, no condensation would form against the wall of your container.

So, if you decide to use open spray foam, you need to install a vapor barrier that is situated between your framing and the drywall. It's not easy to seal everything. Personally, I wouldn't take the risk of using open foam because the chances of rotting are present. You are better of using closed spray foam and paying a little more (if you live in a cold climate).

Closed Cell	Open Cell
Cold climates	Warm climates
Strong and firm	Weaker
Air barrier at 1.5"	Air barrier at 3.5"
Moisture barrier at 1.5"	No moisture barrier
R-6 per inch	R-3.5 per inch
Costs more	Costs less

Open-cell can be beneficial in hot climates, also it has the ability to breathe. So, if there is a leak somewhere, it will evaporate back through the ceiling or wall. It only works if there is a vapor barrier.

Spray foam can also be used as a thin layer (1.5 inches) to stop condensation from forming on the inner walls. If the thin layer has been sprayed on, you can add other types of insulation like rockwool, fiberglass, or others.

This is a table for the recommended R-value in your home, depending on the climate and your heating system. You will note that the R-value for your ceiling is recommended to be the highest. That's because heat transfers are more easily through the ceiling than the walls.

Regional R-Values

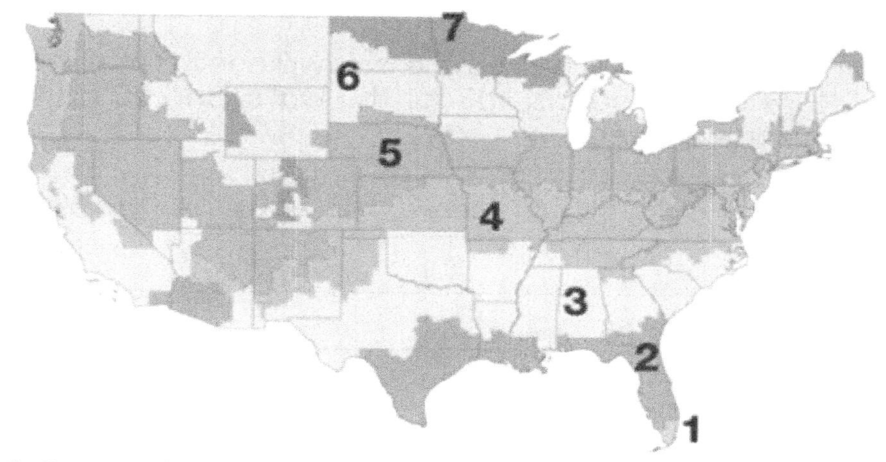

The Department of Energy recommends different insulation levels based on regional climate zones to increase energy efficiency

Zone	Attic	WALLS		Floors	Crawlspaces
		2x4	2x6		
7	R49 to R60	R13 to R15	R19 to R21	R25 - R30	R25 to R30
6	R49 to R60	R13 to R15	R19 to R21	R25 - R30	R25 to R30
5	R38 to R60	R13 to R15	R19 to R21	R25 - R30	R25 to R30
4	R38 to R60	R13 to R15	R19 to R21	R25 - R30	R25 to R30
3	R30 to R60	R13 to R15	R19 to R21	R25	R19 to R25
2	R30 to R49	R13 to R15	R19 to R21	R13	R13 to R19
1	R30 to R49	R13 to R15	R19 to R21	R13	R13

For example, you live in zone 4. You will need a minimum of R-38 for the ceiling, R13 for the walls, and 25 for the floor. You will need:

	R-VALUE (PER INCH)	CEILING	WALLS	FLOORS
FIBERGLASS	3.5	10.9	3.7	
STYROFOAM	5.0	7.6	2.6	5.0
OPEN CELL SPRAY FOAM	3.5	10.9	3.7	
CLOSED CELL SPRAY FOAM	6.0	6.3	2.2	

We will only use Styrofoam as insulation for floors. Read more on this later.

You can use a mix of different types of insulation. You can choose Styrofoam paneling for the roof, spray foam for the walls, and Styrofoam paneling for the floor.

Venting

You can use the existing vents in the container to create airflow. You need to make a box around it, so it's not obstructed by insulation.

If you prefer your own venting system, you need to get rid of these vent's. It's advised to use silicone to cover up the vent. If it's hot outside, but colder on the inside moist air will get inside and create mold and rot if you don't cover these up.

Existing vent holes in the container

Adding Floor Insulation

After you have applied a coating to the floor, framed the interior, installed the electrical, plumbing, and added insulation, it's time to add floor insulation.

While you can add spray foam on the underside of the container, I don't recommend this because I haven't seen anybody do it.

What I do recommend is that you use insulation of about ¾ inch rigid foam insulation and cover it with sub-flooring (wooden panels). Both the insulation and subflooring should be tung and groove.

The reason why we add flooring after the framing is because the frame is mounted on the already existing floor. Which makes it a lot stronger.

Imagine we would screw the framing into the sub-flooring and the rigid foam insulation panels. I won't be very strong. Plus, it would create cold bridges (metal screws), and we don't want that.

After installing the water lines, doing the electrical, and finishing with adding insulation, it's time to add drywall and start painting.

CHAPTER 8
Services

Getting services on your site is not just essential to make it easier for you to live in your container house. You are expected to use items like pumps, water tanks, and port-a-potties without facilities.

Although any organization has to be contacted on its own, resources such as In My Area helps demonstrate which businesses (through different kinds of services) serve your place.

One factor to verify is whether all of the utilities have a minimum monthly fee. If so, we will suggest that you wait until you are able to start building the hook-up. If you do it early, you just have to pay the bill, regardless of when you ever need it.

Depending on where you are, certain services can be deregulated, monopolized easily by a corporation, or governed by the government. If you have an option between several firms providing the same utility service as a customer, make sure to do some homework to figure out which fits better for your case.

You should also verify if they have any energy-saving benefits or rebates that you might have on your template with just a few tweaks. Even utilities provide you with a financial opportunity to use quality insulation and windows, more energy-friendly equipment, etc. Make sure you ask!

Electricity

Electricity is the first and perhaps most significant utility. To figure out how to install an electric meter and link power, you will have to call your nearest electric utility and cooperative.

If you have power lines on the main road near your property now, you should be in a position to add electrical service. The concern as to how much it costs will depend on such factors as installing a new transformer, how long (and difficult) it will be from a run, and whether it will go underground or on poles.

Normally, the firm provides you with a certain gap between the wire and the poles, and you must then compensate the extra distance above that number. You should be able to get an appreciation. Understand that in many situations, you get a discount on the real costs to install it in the belief that you will benefit from your annual subscription charge over time.

This makes it possible for the electrical provider to see some success before they undertake to expand the electrical supply to your site. If they are less than sure that they can ever complete the container home and be a happy client, they will reserve the right to pay for the installation. Or you can pay for a greater share to reduce this chance. Each business is different, so make sure that you figure out what people in your field need.

Understand that permits and permissions are necessary, especially for additional overhead poles that can affect neighbors. You may not have to start the installation immediately on our preceding stage, but you can contact them as soon as possible to clarify the procedure and the timetable involved.

You probably want to get temporary power mounted first as part of this operation, which provides you with some electric circuits. It should be enough to build but too little for the whole building. The business will return later after the completion of the house and have your permanent service mounted.

If you do not find commercial electricity, it might be more appropriate to live off of the grid with a battery, wind turbine, or solar panels than to pay the electricity provider to expand their access to you.

Gas

Gas is perfect for space heating, stoves which water heaters, and usually requires natural gas or propane. If you're in the area, you can reach the natural gas line that you can tap with one meter like electricity.

In more rural areas, you can typically buy or rent a large gas tank that meets your needs month after month.

The easiest thing to find out about the petrol prices in your area is making educated decisions about the types of equipment you choose to have in your home. In general, if gas is affordable, it is the cheapest and simplest to use.

Sewer and Septic

If your property provides access to local sewage pipes, the cost and method of binding must be identified. A septic system is probably your only choice for more remote areas.

The first cost of installing a septic system would typically be much higher than a sewer connection, but after completion, it cost virtually nothing, compared with the monthly cost of your sewage connection.

Most septic systems have buried tanks and lines of leach pipes or sprinklers. Work with your installer to create a good place for this equipment that does not damage future work or livelihood.

Telecommunications

While some people are constructing container houses in rural areas to get away from anything, most people want to have some access at least. The choices available will differ significantly from place to location.

In the area, you will have many ways to bundle tv, Internet, or even telephone services into one bill, including cable, DSL, and fiber. Outside the capital, satellite devices, slower speed cable/DSL links, or even earth-to-point radio frequency equipment may be used to obtain these services.

If you have access to several choices, make sure that you call and match rates and haggle. We would like to talk to neighbors about the solution they use and if they like it.

For example, telecommunications connectivity can be useful to tie up surveillance camera tracking and make Google searches or internet shopping available quickly from the web!

Water

Water is last but not least. The same drinking water you bathe in the United States is also you drink, while in other countries, you will have to prepare to buy drinking water separately.

In any case, all but most remote sites normally have access to sewage. If you can't get affordable access, you would either have to pay to dig a well or to get water trucked into and kept in a tank on-site.

All these solutions have higher initial costs but will be

reasonably rational if you intend to own the container for the coming years.

Security Systems

Security Systems for Shipping Container Homes

Shipping container homes are a surprisingly common sight in many areas of the world. They're inexpensive, durable, and as you'll see here, they make for some pretty awesome security systems.

With the right planning and precautions, your shipping container home should be perfectly safe from burglars or squatters.

- Ensure these buildings have sturdy locks on their entrances and windows for maximum protection against intruders like thieves or squatters.

- Add an alarm system to call attention to potential burglars when they try to get in through the doors or windows without authorization from the owner.

- Install guard dogs as a deterrent for potential intruders like thieves or squatters.

- Add a wide range of security options to your door and window locks, including heavy-duty steel doors, bolts that require a key to be inserted, and security bars.

- Install hidden cameras within the property to capture

criminals in action.

- Make sure the outside walls are solid and secure to prevent burglars from entering through any openings. Make sure they are strong enough to stave off all types of thieves, including those with different types of explosives.

- Consider installing a home alarm system that automatically calls for help in the case of a break-in.

- Add a gas filler cap to your containers. This will allow you to easily check for any traces of gasoline or other flammable liquids around the outside of the container, which could be an indicator that someone has tried to break into your home.

- Make sure every room within your shipping container home is brightly lit to deter intruders from entering, even if they manage to bypass other security measures.

- Install motion sensors and light timers that dim lights automatically when there are no people around inside the premises but still keep them bright enough for security purposes.

CHAPTER 9

Container

Now that we've addressed obtaining your container and preparing the site, it's time to look into shipping the container to your desired building site. An important factor to consider here is whether you'll be making the necessary modifications to convert your container to a dwelling onsite or if the conversion will be done elsewhere.

Other important considerations include the budget for shipping new, used, or one-time use containers and the location from which they are being shipped.

Converting Your Container – On or Offsite?

Converting your container means making the necessary modifications to create the shell of your living space. Certain designs require walls to be removed so that containers can be connected to create larger spaces. Doorways, arches, and windows must also be cut through the steel. Some of these modifications will require equipment, such as welding tools, cutting torches, grinders, heavy-duty drills, and sprayers for insulation. Sometimes, it is more economical to have the containers delivered to a workshop supplied with the necessary tools.

On the other hand, having the containers delivered directly to the site and making the modifications in place can save you an additional shipping cost. The catch is that you need access to the necessary tools and skills to make this option work for you. Here is a breakdown of the pros and cons of onsite and offsite conversion, as well as a third option that might serve you:

Onsite Conversion

If you can manage the conversion onsite, it is helpful for several reasons. The first is that you will have uninterrupted access to the site. You can make your own hours when converting your shipping container to a home. And, perhaps worthy of mentioning first, once you have converted your container, it is already in place and won't require any further travel.

Converting the containers compromise their structural integrity to a certain degree. This isn't an issue if they are already in place; however, if they are converted offsite, then you will have to be doubly careful in transit. Onsite conversion avoids this issue altogether.

There are a few challenges associated with onsite conversion. First, you'll need to have the tools and skills to do the conversion in place. If you intend to purchase the tools specifically for this job, this can be an expensive option. Plus, if you're building on a greenfield site, you'll need to supply power and water. This often means a generator, which can be noisy and expensive. You can also choose to have electric facilities established before placement, though this will have to be planned out ahead of time.

One option that is available to you if you choose to convert your containers in place is to contract a construction team for the conversion process. This sidesteps the need to buy costly tools and makes sure that you have the necessary skills at hand to make the conversion. However, it is a more expensive option than a DIY project. This is an ideal option if you lack the necessary tools and skills, and if it is unfeasible to station your containers at a local workshop to convert them.

A third option is to have your containers converted before delivery. This will be more expensive than either of the prior options; however, you will be able to save quite a bit of time and place your containers directly once they are delivered. Purchasing your containers pre-converted will also ensure that they are both modified and delivered professionally and that they make their way into place intact and without damage.

Offsite Conversion

If you aren't well supplied with tools, then one good option is to have the containers delivered to a local workshop or fabricator. Once they are delivered, you can use the tools supplied there to make all necessary modifications before transporting them to your building site. In addition to the tools, you'll have several experienced people on hand that can help you with any details that require a bit more know-how.

One big advantage is that workshops protect from the rain and the elements. You won't have to worry about making sure your containers are watertight immediately. They will also be protected overnight, so you won't have to worry about tampering or interference when you are unable to be onsite. Finally, while at a workshop, you will have access to water and power, which means you won't have to set up supplies before delivery.

Although offsite modification offers several advantages, there are also a few disadvantages to consider. First, the location of the workshop might be fairly distant from you, requiring a hefty commute to and from it to work on the modifications. Second, the workshop is unlikely to be open at all hours. This means that you may find yourself limited in your access to the container, especially if you are working a regular job in the process. Look into the times that the workshop or fabricator will be open before deciding to use it for your modifications, and try to arrange a flat fee for the month or the week if possible.

Container Placement

Tilting the Container into Place

Once your container reaches the site, there are a few options for placing it. The cheapest option is for the container to be delivered on a flatbed trailer. If the site allows, you may be able to have the driver tilt the bed and slide the container directly onto the foundation. This is also the easiest way to go, and if you can plan your design and site layout to make it a possibility, then you will be able to save the price of renting either a crane or HIAB. To do this, you will need to arrange the foundation so that there is the space of a flatbed trailer and truck adjacent to the narrow end of the foundation, leaving space for the truck to maneuver.

Placing With Crane or HIAB

Sometimes, tilting the container onto the foundation is simply not an option. Also, the tilting option is unfeasible if your construction is more complicated or multi-level. If this is the case, then you will need to lift your container(s) and set it upon the foundation. A HIAB is cheaper, and it will work for smaller containers; however, it may be unable to lift anything more than a 20-ft. container. A crane, though more expensive, will have sufficient lifting power and control for heavier containers and more delicate operations. The typical price for crane rental is USD 700 per day, though this will depend upon the contractor.

Lining and Insulation

One important tip when placing your containers upon the foundation is to line them with a polyethylene damp-proof membrane. Also, if you use a crane or HIAB, you will have access to the underside of the containers. You can lift them one by one, sandblast the bottom, and add a 1 in.

coating of polyurethane spray-foam insulation. This will reduce heat loss from the bottom of the container. Even without the benefit of a crane or HIAB, you can treat the underside of the container if using a concrete pier foundation.

Tip: If your foundation isn't completely level, you may need to use shims, metal spacers, to raise the container and bring it level.

Tip: By spraying foam insulation between any connecting walls once the containers have been lined up, you can keep moisture out, reduce drafts, and help maintain the internal temperature of the containers.

Stabilizing Your Containers

Cleaning the Containers

Once you have placed your containers, you'll begin to see the shell of your home shaping up. The next step is to connect them. First, however, they will need to be thoroughly cleaned. This will be more important for used containers than new or one-time use. The sandblaster and pressure washer are the quickest options for cleaning; however, in a pinch, you can use a grinder or even wire wool. Make sure to clean the inside of the container, including the wooden flooring. Then proceed to the outer walls and roof of the container.

Stabilizing the Containers on the Foundation

In most cases, the weight of the container alone will be sufficient to seat the container firmly upon the foundation. However, if you have chosen to place steel plates into the surface of the concrete on the corners of the foundation, the containers can be welded to the plates to stabilize them further.

Another option is to bolt the containers to the foundation. To do this, you will want to drill through the bottom corner fittings of the container into the piers, piles, trench, or slab. Once the hole has been drilled, you can place a 12-in. by 1-in. bolt through the hole. Make sure that you use a washer around the head of the bolt to seat it firmly to the bottom of the container. Then hammer the bolt into the hole you have drilled. To achieve the final snug, tighten the head of the bolt. You will only need one bolt in the corner of each container to ensure that they are solid and secure.

Join More Containers

Once the containers have been placed and bolted or welded to the foundation, it's time to connect them securely to one another. You have three options at this point: bolt, weld, or clamp.

Clamping

The least secure (and least expensive) of the three is to clamp the containers together. It does offer the option of disconnecting the containers from one another in the future, should that be desired. However, given that they have already been bolted securely in place, other options should be used if at all possible.

Bolting

Bolting the containers together is the next option. It is more secure than clamping and only slightly more expensive. If you opt for this option, the containers should be bolted together at the adjacent corners. Drill through the corner fitting points from one container to the next. You will also want to drill through a metal plate. This will act as a washer for the threaded side of the bolt. Insert the bolt through the hole (including a washer), slide it through the containers, and then place the metal plate, an additional washer, and a nut on the threaded end. Torque the nut tight, and then seal any gaps by placing mastic around both ends of the bolt.

Welding

Like clamping, bolting leaves the option of disassembling the containers later should the need require. However, the best option by far is welding. Welding makes the overall structure more rigid and secure. It also helps to keep the containers level despite settling. If you have access to the equipment and tools, then welding is by far the best option for a long-lasting shipping container home with a minimum of repairs needed over its lifetime.

The containers should be welded at the jointure of the roof, floor, and end walls. One of the best methods is to place a 3-in. x 1/8-in. length of flat steel against the jointure of the roofs and secure it with a stitch weld. Once this has been welded in place, repeat the process for each end wall with a 2-in. x 1/8-in. length of flat steel. Finally, use another 2-in. x 1/8-in. piece of flat steel to weld together any overlapping floors of adjoining containers. This will ensure that all of the contact points between containers are welded securely to one another.

Tip: To prevent rust, place a few layers of latex paint over each of the flat steel bars. Make sure to completely cover the area of the weld.

Cleaning the Container

Even though building or owning a shipping container home is cheaper than subscribing to pay a mortgage for thirty-plus years of your life, it's still no small drop in the bucket. A shipping container home is an investment, and your container needs to be taken care of to ensure that it has the longest life possible. This will focus on steps that you can take to lengthen the life of your shipping container. The last thing that any new homeowner wants, a shipping container or a standard two-story home, is to purchase something only for it to deteriorate and falter under environmental and other types of pressure. This will focus on what you can do as a responsible homeowner to reduce the wear and tear of your home over time. With these tactics in use, you'll greatly increase the overall and long-term value of your home.

Tactic 1: External Cladding

You can think of external cladding to be the equivalent of a protective layer similar to a skin or coating. This will protect the outside of your home from weather and other factors that will wear it down over time. Cladding does not remove the original steel exterior of the shipping container home but instead pads it. Additionally, external cladding can serve to make the steel quality of your home appear less industrial and softer. The types of cladding that you can purchase to both weatherproof and improve the outer look of your home include:

- Stone cladding
- Timber cladding
- Weatherboard cladding
- Brick cladding
- Fiber Cement cladding

Adding external cladding to your home will ensure that termites and rot cannot get to it. External cladding can last well over fifty years. You can find cladding for as little as $3.00 per piece to as much as $30.00 per piece. The vast range in price can give you many possibilities and choices.

Tactic 2: Treat Areas of Rust as Quickly as Possible

Rust can damage the overall quality of a shipping container home, so it should be no surprise that if you treat areas of rust as quickly as possible, you can avoid having to make pricey decisions for your home at a later time. While it's possible to remove rust using ingredients from your home, such as lemon, this method is not advisable to be used on something as valuable as your container house. Instead, try the following combination of products if you want to see maximum results against any rust that has accumulated:

- 7 cups lime-free glycerin
- 1 cup sodium citrate. This product can be found at the drugstore
- 6 cups lukewarm water
- Powdered calcium carbonate (also known as chalk).

Keep adding chalk as needed until a paste forms. After you have your pasty mixture, spread the paste on the area stained with rust through a spreading tool of your choice and wait for it until it hardens. Once hard, using a metal tool, like a chisel or prong, to scrape away the residue from the mixture that you have made. This should remove the rust that was once on the house, but if it doesn't, repeat this process until it's gone. Remember, as a container homeowner, you should be looking for rust regularly because this is one of the problems that frequently occur in these types of homes.

Additionally, it's important to realize that when the shipping container is transported to you, it's likely that there will be dents in the roof because of how the shipping of these units usually occurs. Be sure to get the dents out of your roof when you start building your first home. This will save you trouble in the future.

Tactic 3: Use Corrosion Resistant Paint

Another method that can serve as a form of maintenance for your container home is to apply corrosion-resistant paint to your unit. If you use this method of protection, consider using it when your home is first being built so that you can prevent future corrosion. You can use this type of paint after an area of your unit has already started to rust, but this is less of a preventative measure before the incident and more of an attempt to fix something after it has already occurred. Take the time to use the precaution of corrosion-resistant paint when your home is first built so that you can avoid the pain that rust brings to the eye when you see it on your shiny new living space.

Tactic 4: Grease

Another tactic that seems to be more preventative is using grease in areas prone to sticking or rusting over time. These areas include the door hinges of the shipping container unit (if you decide to keep them), window jambs, and door handles.

Because shipping containers last decades being transported internationally and overseas, we know that they are relatively long-lasting and durable; however, because people have not been using them as homes for very long, it's hard to know exactly how long they last and what their great limitations seem to be. Of course, the most current and significant limitation in terms of maintenance seems to be rust. The accumulation of rust should be avoided whenever possible. When rust does occur, the solvent that was presented will definitely help to eliminate what rust has collected. As with any type of injury, if a significant problem can be avoided by small and frequent measures, performing these measures is the preferable option to avoid a large problem in the future.

CHAPTER 10
Roofs

Roofing and installation choices are just as important as any other aspect of preparing your shipping container home before moving in. Choices may be limited depending on geographic location.

As always recommended, check with your local building department to make sure that your choice is not only legal, but appropriate for your city, county, or state ordinances.

In some places, you may decide to simply go with the metal top of your shipping container, especially if your container placement is only temporary. However, do be aware that leaving the steel roof unprotected from the environment can accelerate corrosion and/or possible damage depending on climate scenarios.

Roofing your shipping container home comes with several pros and cons.

Pros:

- Protects the roof from inclement weather
- Aids in insulating against cold or heat
- Protects the steel container from puddling water and helps to protect against rust and corrosion

Cons:

- More cost outlay

- The necessity to adhere to local building codes

- If you've decided to go with a roof, your next step will be to decide which type!

Roof types

Suitable roof types are suitable for shipping container homes:

- Flat roof

- Sloped roof (often called a shed roof)

- Raised or gable roof (will require trusses - not typical but an option)

- "Green" roof - traditional with pioneers and early American settlers, sod or earthen roofs are also an option

Carefully consider each option depending on aesthetics, cost, and local building regulations.

Individuals looking to move into a shipping container home are often ecological-minded. For those, a green roof or earthen roof provides several benefits.

They enhance installation capabilities, reducing heating and cooling costs. They're natural and aesthetically pleasing, relatively low maintenance, and they're fun!

Before deciding on a green/earthen roof, however, be aware that you'll need to waterproof the ceiling of the shipping container with a membrane or other roofing material to protect the structure against damage.

Tip: Remember that your shipping container also needs to be strong enough to support a sod roof! Consult with a building or roofing expert to determine the load limitations of your steel container depending on size.

Every type of roof construction must also follow local building regulations concerning slope or pitch.

Preparing For A "Green" Roof

Several steps are necessary before you simply cut out chunks of grass and place them on top of your shipping container. First, verify slope specifications with your local building department.

You'll need to follow a few steps before you 'plant' your green roof:

- Some type of waterproof membrane that will offer protection to the top surface of the shipping container.

- Sheeting that will act as a root barrier. Polyethylene is one choice.

- Some type of aggregate (pebbles or mesh, for example) to enhance drainage when it rains.

- Some type of filtering fabric such as burlap or

weed barrier paper. This is common in landscaping scenarios to prevent weed growth. Weed barrier paper or fabric blocks the sun, preventing the growth of weeds. Be aware of the difference between landscaping fabric and filtering fabric. Both are geotextiles, but landscaping fabric serves as a barrier while draining fabric serves as a filter.

- Planting soil/fertilizer/nutrient-rich dirt that will provide nutritional support for sod.

- Grass, low-growing and shallow root groundcover, or other choices

Flat Roofing

Flat roof shingles or asphalt shingle rolls are an option for covering the flat top surface of the shipping container. Application methods will depend on the type of materials chosen.

Tip: A flat roof is not conducive to several geographic locations around the US, such as those that receive heavy snowfall in the winter months.

Sloped Or shed/ Angled Roofing

Sloped or shed roofing is recommended in areas that receive a lot of snow in the winter. They're also conducive to the installation of solar panels. Pitch requirements may be quite specific depending on your geographical location.

Regardless of the pitch or slope of the roof, choose materials that will not only provide the insulation factor and protection you'll need depending on the environment, but that will not only match the ultimate plans for the exterior of your home with local ordinances and preferences.

For example:

Asphalt shingles (wood shingles are rarely used in the US anymore due to potential fire dangers)

Asphalt shingle role (tarpaper, adhesive, and easy roll-out application makes for easy, fast installation)

Galvanized or metal roofing (coated steel) is preferable in areas that get a lot of snow.

Points to remember:

- No matter what type of roof you choose, opt for one that is structurally sound, suited for your environment, and that your shipping container (based on size) can easily bear.

- Green roofs are beautiful, ecological, and low maintenance, but they are also extremely heavy. Be aware that additional steel support columns may be required inside the shipping container to evenly distribute the weight.

Tip: Before designing or deciding on a roof, refer to the expertise of a structural engineer to expertly provide calculations for support and load limitations of your shipping container.

CHAPTER 11
Flooring

Moving on to flooring is an exciting time. You'll begin to see your home really taking shape at this point. And like with every step of the process, you have a few options. Basically, whichever way you choose to go about it, there are two steps: deal with the original flooring and lay the finished flooring.

When your container arrives, it will come equipped with a marine plywood floor. You may be tempted to just stick with the existing flooring, but you'll have to keep a few things in mind. This floor is treated with pesticides and other hazardous chemicals. These chemicals will leach into the interior of your home and create health hazards. So, the floor should either be removed and replaced or sealed before moving on. If you are working with a brand-new container, you have one more option. You can actually arrange for the container to be delivered without flooring. This will save a step in the flooring process, but remember that new containers are significantly more expensive and less environmentally friendly. The final option is to go with prefab containers. In this case, you won't have to worry about the flooring at all, as it will be taken care of for you.

Another thing to consider here is that you may wish to replace the flooring, set up a subfloor now, and then do the final floor after framing and finishing the interior. You'll have to set the construction schedule based on the flooring type you choose. These options include creating new flooring, using a non-breathable underlay, or pouring a concrete floor.

Let's take a closer look at the process and your options along the way.

Removing and replacing the Existing Flooring

One thing that you'll want to keep in mind when sourcing your container is to check into the condition of the existing flooring. If you can examine it yourself, you can inspect the floor for holes, dents, cracks, or any type of damage. If you are ordering from a bit of a distance, you'll have to rely on the supplier's inspection, but it helps to know what to ask. Suppliers are more likely to forget to mention a detail like that than lie outright. If the existing flooring is damaged, then you may not have another option aside from removal and replacement. Remember that if you choose to go this route, you'll need to factor the price of the plywood into your budget.

Removing the floor isn't complicated, but it'll take a bit of time and sweat to make it happen. And if you're working with two or more containers, expect to spend significantly more effort on this step of the process. After removal, then new plywood should be cut, laid, and fixed into place with self-tapping screws. Keep in mind that it's best to handle the flooring before framing the interior with stud walls.

Installing a Subfloor

Another option, if the existing flooring is not damaged and you don't want to remove it, is to install a subfloor. The entire purpose of this stage is to make sure that the hazardous chemicals in the existing floor aren't seeping into your living spaces. So, when laying a subfloor, the first step is to seal the existing floor.

To seal the floor, begin by cleaning it thoroughly with isopropyl alcohol. After it's been cleaned, coat it with a low-viscosity epoxy. Low v epoxy works great in damp conditions and when you're dealing with high moisture levels. To plan out how much you'll need, remember that low v epoxy is sold in 1.5-gallon kits. Each kit will cover from 150 to 175 sq. ft. of flooring. One coat will seal the flooring in most instances, but if you'd like to be extra careful, you can use two. The epoxy will create a vapor-proof barrier that effectively contains the hazardous chemicals in the original flooring. Within eight hours after placement, the epoxy will have set enough to handle heavy traffic. A tip for those doing this the first time: Start in the corner of the room and work your way back to the door.

After the original floor has been sealed with epoxy, the next step is to lay down a layer of plywood. Three-quarter-inch marine plywood will do the job perfectly. You'll also make it easier on yourself if you go for tongue and groove sheets that can be secured to one another while placing them. After placing a sheet, drill through it and into the original flooring with 2" coated deck screws. Before placing your plywood subfloor, you may opt to place a half-inch foam layer down over the sealed flooring. This will provide more insulation, though it'll take up a bit more height.

The issue of height is the only downside to laying a subfloor. You'll be losing an inch from the finished height of the interior space. Since interior space is at a premium, every inch counts. Another consideration is that it is just as expensive to install a subfloor as to remove the flooring and replace it entirely. You'll be saving time, but not much in the way of money.

Non-Breathable Underlay

If you are looking at the absolute cheapest and easiest option, then you can choose to place a non-breathable underlay before finishing the floor. This is a very simple process and takes no more than a couple of hours for a single 40 ft. container. Another advantage to this approach is that the bulk of the work can be done after the interior is framed, just before installing the finished flooring.

To install an overlay, you'll need to begin just as you did for installing a subfloor. Clean the existing flooring thoroughly with isopropyl alcohol and seal it with one or two coats of low v epoxy. You can then either carry on with framing and finishing the interior walls, or place the underlay directly after sealing with epoxy. In either case, measure out the space you're working with. Cut the underlay to match the dimensions of the floor, then line the underlay up on the floor as needed before nailing it into place. This can be done while installing the finished flooring. This will protect the underlay from damage and secure it in place in one simple step.

Concrete Flooring

Concrete is the last option for dealing with the original flooring. If you go this route, you won't need to seal the floor, place an underlay, or install a subfloor. You can pour concrete directly on the existing flooring, and it will create a natural sealing layer. Plus, you can use the concrete as your finished flooring, saving yourself a few steps later in the game.

Concrete is great for the floor in several ways. It's simple to clean, strong enough to last, and can be dressed up in many ways. The floor can be dyed to add a bit of color, polished for a sparkling finish, or set up with patterns or textures. At the same time, there are a few disadvantages. Concrete will absorb the cold during the winter so it can be a bit chilly on the feet and you'll have to pay a bit more in heating costs. Plus, you'll need to put steel reinforcement in place to provide flexural strength.

The easiest way to set up the steel for your flooring is to weld 2 mm steel bars across the length and width of your flooring. Set them at about an inch above the plywood, spaced a foot apart from one another. After welding them in place, it's time to pour! Once again, work from the corners to the door, finishing the concrete as you go. The floor should be about three inches thick when finished. It'll help to mark the finished height before beginning the pour.

Finishing the Flooring

For the options described above, you'll want to make sure they're done before framing and dressing the interior. If you have chosen to go with concrete, this can be used as finished flooring. However, the other options will require a flooring finish that looks good, has the qualities you want, and fits the design of your interior.

So, what are your options here? Aside from concrete, you can finish the floor with tile, carpet, or laminate. You can even install hardwood flooring if you like the look and want to go that extra mile. The biggest thing to consider when choosing your flooring is temperature. If you live in a warm climate, you'll want to go for something that helps your home to cool. Concrete, tile, and laminate are all good choices here. They all hang on to the cool temperature and help to keep your space comfortable when outside temperatures are skyrocketing. Cold climates and severe winters will benefit from carpeting. This is more of a challenge to clean, but it feels a lot better on the feet when it's the dead of winter. It will also hold on to the warmth in your space, cutting down on heating costs.

Whichever option you choose, you'll have to start by measuring out the floor. This will let you know what to buy in terms of materials. Get the total square footage of the floor space. If you're doing this step after framing and finishing the interior, you'll need to take note of odd-shaped spaces and the dimensions of each room, keeping them in mind when buying and cutting carpet or vinyl.

Laying Carpet

If you live in a place that gets cold in the wintertime, and you like to walk around your home barefoot, or you just like the look and feel of carpet, then the carpet is the way to go. Although it's a bit more challenging to clean than the other options, it's also relatively easy to place.

When laying carpet, you first have to prepare the space with carpet grippers. These are thin strips of wood with sharp pins coming out of one side. They are used to hold the edges of the carpet in place. Line the edges of the rooms with the carpet gripper and nail them into the flooring or fix them in place with carpet gripper adhesive. Do this for all the rooms where you intend to place the carpet. Leave about 10 mm of space between the carpet gripper and the wall itself.

If you've worked with carpet before, then you know that the next step is to place a carpet underlay. This will provide a bit of cushion, making the carpet more comfortable to walk on. The carpet underlay has one rubber side and one foam side. Place the rubber side down against one wall on top of the carpet grippers. Unroll it until you reach the opposite wall, and then cut off the excess of the underlay with a utility knife. Snug the underlay against the carpet gripper. Keep on going with this until the entire space has been covered with underlay. The jointures between one layer of overlay and another should be sealed with carpet tape before moving on.

After the underlay is in place, the next step is to lay out the carpet. You'll want to cut out the desired dimensions of the carpet so that you can fit them within each room. Most of the time, you'll be cutting the carpet from the back. If you happen to be working with an oddly-shaped room, then remember to reverse all your measurements so that the carpet will fit in place.

So, measure twice, cut once. After you have carpet in the right dimensions, you can bring it into the room and lay it into place loosely. Start with one corner and make sure that the carpet is securely fixed to the carpet gripper. Leave about 50 mm (2 inches) of space, as the carpet will stretch a bit as you get it into place. After securing one corner, move along the wall to the next corner, stretching and fitting the carpet to the gripper all the way around. Cutaway all excess material with a utility knife so that the carpet fits the room perfectly. Also, remember to work from the far wall to the door so that you can easily remove all excess material at the threshold.

CHAPTER 12

Framing and Ceiling

Is an Interior Frame Out a Must-Have?

Container home builders with a keen eye on the bottom line and a desire for a move-in ready home as soon as possible tend to skip the step of framing out their container's interior walls

Others actually prefer a rough, unfinished look that clearly shows off their home's unique origins, so they leave their interiors frameless for purely aesthetic reasons.

However, if time and additional costs are not an issue and you want a container home that looks nearly identical to a "regular" home on the inside, you may want to create a finished look by installing a wallboard. This step will require that you frame out your container home's interior.

The reason that a frame out is necessary in such a case is that, as you install your wall boarding, you will be drilling into the frame that you put up rather than directly into your actual container, a disastrous move that will leave you with a pocked box full of numerous holes.

Another case was framing out the home's interior is necessary is when you will be making use of an insulation method that requires the use of stud walls. Such insulation methods include fiberglass batting insulation, natural cotton or wool blanket insulation, and insulation panels, all of which need to be applied to frame stud walls.

However, if you've chosen good quality closed-cell polyurethane spray foam as your home's insulation method and you've decided against wallboard, you're home-free when it comes to giving the step of interior framing a miss.

If you do decide that you would like to proceed with framing out your container's interior walls, keep in mind that the frame out process is very similar to what you would be doing inside a "regular" home, except for how you attach the framing to your container's walls.

Some things to consider before beginning include:

Steel Studs or Traditional Wood Studs?

There is a growing move towards using steel studs in place of the usual wood studs when framing the interior walls of container homes. This is due to the idea that wood studs are far thicker and will take up badly needed interior space, whereas steel studs are more compact and will provide a flatter frame out, giving you more room to enjoy inside your container home. However, having been involved in builds where steel studs and wooden studs have been used, I have gotten to see, firsthand, the difference that it makes. And I can confidently report that the difference is very slight indeed.

In fact, when all is said and done, steel studs will really only give you about 2 inches of extra space on both sides, working out to an additional floor space of approximately 6 square feet, in most cases.

Another downside to using steel studs for interior framing is that they require more work and take far longer to complete than using wood studs because each steel stud must be clamped and screwed in 4 places.

And the final, and perhaps most important reason I don't recommend using steel studs in place of traditional wood studs in most cases is that they cost a lot more. One of the main benefits of building a container home is that you have a super-strong steel structure with all 4 walls and a ceiling that needs low to no reinforcement for most designs. This saves home builders a lot of money and effort, but when you have such a structure already in place, paying more money just to add more steel to frame the interior simply to save an inch or 2 here and there doesn't make sense to me.

The only case where I would recommend you go with steel studs over traditional wood studs is when you are putting up your container home in a moist, warm tropical environment such as Costa Rica. I know that many of the container homes built in Costa Rica and other similar climates use steel stud framing welded as a way to combat the moisture-heavy air, frequent tropical rains, and the abundance of insects that exist in such areas. If you are putting up your container home in Southern Florida, for instance, I'd say, take a look at steel stud framing as an option, simply to keep any concerns about the bad combo of humid conditions and wood at bay.

If you do need to go with steel studs for climate concerns, remember that as steel is such an excellent conductor of heat, you will want to insert a thin sheath of foam between the upper box beam. This will effectively produce a thermal break that will prevent the loss of heat through the container walls and also prevent condensation from the studs affecting your drywall at a later stage. This isn't something you'll need to worry about with wood studs, however.

If a tropical climate isn't a concern, then, by all means, reach for regular wood studs. They'll save you money on paying for the more costly materials and any additional labor, as well as time and effort, and give you nearly identical results for a lot less!

When doing an interior frame out with traditional wood studs, you'll likely need little help. Even the most basic rough framing will work very well, and anyone who's created a wood frame of any kind in the past will be able to achieve a good frame out. Using wood 2 by 3's or 2 by 4's and a power framing nailer, create your framing. Then you can attach it to the container by screwing the bottom plate into your container's floor, using wood screws. Wood screws should be coated or treated to withstand any chemicals that may exist in your flooring. The upper plate of your frame should be screwed into your container's upper box beam. I recommend using metal thread cutting screws for this. Be careful not to allow your screws to reach and go through your container's outer shell as you work. Grabber caulk can also be used with success.

One thing to remember when working with wood is that wood tends to shrink over time. The best way to combat this is to ensure that you are working with extremely dry wood. Otherwise, any existing moisture will exacerbate shrinkage problems and affect the integrity of your frame out.

Money-Saving Tip

Another emerging method of finishing the interior walls of container homes that is fast becoming popular these days is to actually use an adhesive and stick insulated board directly to the container's interior walls. This saves the home builder a lot of money and can be managed alone. One caveat, however, is that it is more labor-intensive, but if you'd rather work a little harder and spend a lot less, this is definitely a method to consider.

With this relatively new method, I would advise that you use a stiff foam board and a good, strong adhesive to glue this board directly to your container's metal walls. Because moisture always finds a way in if not properly prevented against, I would caution that you take the additional step of spraying foam around the entire perimeter of the board to stop dampness from collecting behind it. Without this step, you could be creating a moisture trap, so don't skip it if you decide to try this money-saving method.

CHAPTER 13
Interior Design Ideas

Now that you are almost finished with the interior of your shipping container home, maybe it is time to come up with design ideas to make the place look and feel a little bit bigger. While you can add more than one shipping container to maximize your floor space, that is not always an option due to budget restrictions, and sometimes, you just have to make do with what you have. The default for shipping containers is that they are snug, which is why every step you take needs to be calculated, whether it is the style of furniture you will use or the location of that furniture.

In this next part, we will explore design ideas to help you make the most out of the space you have without ever feeling it is too tight. We will also talk about suitable furniture pieces and much more.

Mezzanine Floor

Installing a mezzanine floor is one of the best ideas you can make in your shipping container home to save space. It is basically an intermediate floor between the ground and your ceiling, on which you can play around with design ideas to save space. For instance, some people add the intermediate mezzanine floor and turn it into a living space, into which you can add a sofa, hang your TV on the opposite wall, and just chill in your home. Underneath the mezzanine floor, you can put the bed and turn it into the bedroom.

Sure, this takes up a lot of the headspace in a shipping container, but it also saves you a lot of floor space you can otherwise use cleverly however you please. The mezzanine floor in our example allows you to have the bedroom underneath the living room all in the same space. You can even play around with the furniture you use to make the most of the space, so the sofa on the upper floor can fold into a bed, and so on.

Add a Pegboard

Pegboards are a pretty useful item to have around any house, especially a shipping container one. They are excellent for storing smaller items that would normally take up too much closet and drawer space. You can store all your cooking utensils like pans, spatulas, knives, sieves, mashers, and much more if you add the pegboard in the kitchen. Maybe you need space for your art room, so you can hang your brushes, drawing pens, tapes, scissors, or anything you use for your art projects.

Using a pegboard clears up space on desks, in drawers, and around the entire house. You just attach them to the wall, add hooks to the pegboard, and you can start using them as your personal storage space however you please. A pegboard can also be used in the garage to store tools and any other place in the house. The great thing about pegboards is that they come in different sizes. You just need to figure out the estimate of the size you need it in, depending on what items you will store there, and then you can go get one of standard size or more than one and use them together.

Use Dead Spaces

There often are quite a few dead spaces around an average house, whether it is a shipping container home or a traditional one. With a touch of creativity, those spaces can be used to great effect and they can help you save a lot of space. You can start with the space under your bed. Many people add the bed and are done with it, but the average bed takes up a lot of usable space, which is quite valuable in a relatively small space like shipping containers.

This is why you need to take advantage of that space. There are different ways to go about this. Many beds come with dedicated frames that already contain drawers underneath the mattress, and getting one of those can prove useful. This could even save you the need for a wardrobe in the bedroom because you basically have it underneath your bed, so you could store clothes there.

Another space often underused in houses is under the stairs. If you stacked shipping containers in your design, this means you have at least a second floor, and there will be stairs leading to that floor. There is always space under such stairs to be used if you add a few shelves and get a bit creative. You can store books, kitchen utensils, or anything that needs storing around the house.

These are just examples of spaces around the house that can be utilized if you take the time to consider them. If you think hard enough, you are bound to find others that can be leveraged to a greater extent.

Fold

One of the smartest, and coolest, ways you can save space in a shipping container home is to fold everything and anything that can be folded. This might seem like a stretch to you, but it has been done often before and many people have invested time and effort in such approaches and they ended up with wonderful homes with properly utilized spaces.

You can basically fold, swivel, or rotate everything in your home when you are not using it – if you design it right. Your bed can be pulled out of a wall and become a full-sized bed. You can have your entire kitchen hidden behind wooden panels that can be folded open. The same goes for storage spaces and sofas. Anything can be folded with the right design. It might not be perfect for everyone, but it can certainly save a ton of space.

Install Pocket Doors

To save space in your shipping container home, you must get creative with your interior design ideas, there's no way around it. A great way to play around with the design is by installing a pocket door. A pocket door is similar to a barn door, but the great thing about it is how the door retracts into a hidden space (pocket) in the wall instead of swinging open into the room. This means that the whole wall will be free to use when the door is open.

So, there won't be a door taking up space on either side of the wall, and you can get creative with the empty space, too. You can add furniture adjacent to the door, shelves, or anything. This is why pocket doors are ideal for shipping container homes as they can help you maximize your space.

If you think you have the space for it, you can also install a barn door. It has its own rustic, unique design that looks like an actual barn door but knows that the track holding the barn door is mounted on the wall, so it does take up space, unlike pocket doors. Empty wall spaces prove useful, so unless you like the elegant design of that barn door so much, pocket doors are better.

Furniture

The furniture in your shipping container home will play a huge role, not just in making the place look nice, but also in saving space. Every piece you get has to be with careful consideration to the space of your home and whether the furniture piece will help save and utilize space. Invest in multitasking furniture that can serve more than one purpose around the place, thus helping you maximize your floor space.

A great example of multi-purpose furniture is a couch that converts into a bed. If you happen to be living alone in a single-container home, for example, such a couch will be of great use and can save you a lot of space and maximize the living area. Your living room and bedroom could be the same, where you just fold and unfold the couch to change it from a bed into a couch or vice versa. Even if you live with others, a foldable couch is quite useful and can accommodate guests and help make room for many things. Here are different types of convertible couches.

Chaise

A chaise is one of the most elegant additions to any home, and it is very useful in tight spaces despite looking somewhat large. Translated into a 'long chair,' a chaise is an extended chair that provides support for your legs, but it is a one-person chair, so it has a short backrest. It might not be perfect for overnight sleep, but it is excellent to rest for a bit or watch TV.

Futon

It not only looks exotic and classy, probably due to its Japanese style, and it also is quite functional. A futon is a mattress that folds across its length, and when you fold it up, it is a sofa. So, you fold it when you want to sleep and then unfold it to an 'L' shape when you want a sofa.

Foldout Couch

At first glance, the foldout couch (also known as sleeper sofa) looks like a traditional couch, but there is one major difference. When you remove the cushions of a sleeper sofa, you will find a folded mattress and a frame that can be pulled out to form a bed. The cool thing here is how the cushion you sleep on will differ from the surface of the sofa you sit on.

Daybed

Daybeds look quite elegant in any house, and they also are very functional. The daybed has a regular mattress for sleeping, which has a headboard that you pull out to make the bed. When used as a couch, the headboard is the backrest.

Leveraging the Walls

One thing you must do to make the most out of your shipping container space is leveraging the walls. We mentioned earlier that you could hang pegboards to utilize empty walls, but there is much more that you can do. You don't have a ton of floor space in a shipping container home to install closets and large cabinets, but you have a lot of overlooked spaces on the walls you can use. Your walls are assets as much as your floor space.

If you happen to do a lot of work from home, then you might need a home office. Needless to say, getting a large desk is not practical and can take up a lot of space. Fortunately, there is an alternative. You can mount a foldable desk to a wall, which is one of the best ways to utilize space in tighter locations. You can use your desk for work or whatever, and after you are done, it folds flat against the wall, saving you room to move around or do anything you want. A foldable desk uses brackets or legs to support it from the underside and wires or chains from the top. You can make use of the space above the desk, too, so you can add a few shelves and put flowers, books, or whatever you want to decorate the area.

Another great use of the walls is adding TV mounts. If

you're getting a TV, then it could be sizable, and resting it on a table would just be too space-consuming. Consider other options. You can mount the TV on the wall, and like that, you don't need a table or a stand. The great thing about TV mounts is that they come with a lot of options, and some allow you to adjust the viewing angle as you please so you could view the TV from several angles. In other words, you can use the same TV for more than one space, which saves both time and space around the house.

Last but not least, fans mounted on the wall are another great way to leverage the walls inside your shipping container home. While pedestal fans are great, they take up floor space. Ceiling fans, on the other hand, take up ceiling height in an already limited space. This is why a better choice would be mounting the fans on walls. This gives you the benefits of both ceiling fans and free-standing ones. This naturally saves up space around the living container and gives you exactly the type of cooling you need.

CHAPTER 14

Interior

This part of the project is where you get your hands dirty, and you'll need all the tools you can get, from plasma torches to cutting discs. It would be a good idea to consult with a structural engineer at this point as well to ask about what walls can be removed and what to avoid to preserve the structural integrity of the shipping container.

So, where do you need to start in transforming the container?

Adjoining Containers

One of the most common changes people make when converting shipping container homes is to open up adjoining containers to increase floor space or make more room. Many people think that working with metal is difficult or impossible, but it is not if you have the right tools and know what to do. In any case, you have to cut through a lot of walls to make room for doors and windows and open containers. Cutting metal may seem daunting, but it's actually easy because, if you do it right, the results can be extraordinary and clean. It is a material that can be shaped into any shape you want, which is very useful.

Eliminating the walls between two adjoining containers only seems logical to increase the living space and turn it into one large container, considering the small space inside a single shipping container. The first thing to do is to mark and measure the walls to be removed if multiple containers are involved. It's also a good idea to consider what you want to do with the doors before working on the walls. Let's say you have three garbage cans together. For example, will you leave the three doors intact? Or maybe you want to incorporate them into the design in some way. Your last option would be to weld them together and treat them as a regular wall.

Installation of an Auxiliary Container Joining Structure

The trick to working correctly in side-by-side containers is to place them side-by-side exactly as you want them to be. If the conversion is done off-site, this would be the only concern you have and the only thing you have to get right. But if the conversion is done on-site, it's a different story. Remember to make sure the containers are securely connected to each other with bolts, welds or clamps before you get to work on the adjoining walls. Next, get to work with a cut-off wheel or plasma torch and remove the wall space you want to eliminate. It may be the whole wall or just part of it, depending on your design, so measure and mark the parts you want to cut out before you get to work.

If you are not going to cut the entire wall, it would be wise to line the adjoining sides of the container with spray foam so that the remaining parts of the wall have insulation. After cutting the walls, weld steel plates in the gaps between the two openings to ensure the structural integrity of the container walls, but not after spraying the insulation if you haven't already done so. Next, just join the pieces together and finish.

If the conversion were to be done off-site, you would just need to make sure the openings are aligned and the interior walls are aligned. Having the plans clear from the beginning will help you with this part, as you will know exactly where the removed walls will be and how to align the containers accurately. Whether the conversion is done on-site or off-site, double check measurements and markings, as this is an area where incorrect measurements can be problematic. Also, double-check sheet steel connections between adjoining walls and roofs and make sure there are no loose pieces here or there, as they can jeopardize the structural integrity of the entire house.

Pro tip: Wear protective gear, including gloves, goggles and/or face shields if possible. Cut metal is sharp and can be dangerous, and at this stage you will have your hand near it often, so you should take safety precautions.

Floors

Some people forget to weld the floors after they're done with cutting through the walls. Just as you welded the remainder of the walls with steel plates, you need to do the same with the floors to turn the multiple containers into just one unit—the last thing you need is to walk into your house after it is done to find gaps between the floors. Aesthetics aside, welding the floors together also strengthens your shipping container home's structural integrity and naturally eliminates the chances of any pests sneaking in from the floors or leaks happening.

One last very important detail here is the structural reinforcement if you plan on removing large chunks of the adjoining walls. In that case, you need to use steel box beams to support the load coming from the roof and ceiling, and they need to run across the width of the containers in which you have made those large cuts. Stitch-weld the steel beams to the interior of the container roof. As always, consult with a structural engineer here to tell you exactly the bearing loads that need to be supported by the beams so you can understand which kind to get and of what dimensions.

The great thing about dealing with adjoining walls is that you have a lot of options here. Many container owners like to create arches between the two adjoining containers, which increases the space and gives an elegant look and the illusion that the surface is larger than it is. If archways aren't your thing, you can separate the wall between adjoining containers into segments, with some leading to rooms and others leading to a shared living space, for instance. The possibilities are endless, and making changes through adjoining walls is easy enough to give you that flexibility.

Doors and Windows

With the walls now open between the containers, your

shipping container home should be taking shape quite nicely. Next, you need to start working on the doors and windows and the frames you need for each. The first step to working on windows and doors is taking their measurements and marking their locations on the walls. As always, these measurements need to be accurate, and they have to be double-checked because you will be changing the frame of the container, so you can't afford to make any mistakes here. Some experts recommend using cardboard templates for all the doors and windows you will be working on and marking those. Get it right the first time.

Next, cut the walls of the container following the measurements you have taken for the doors and windows. Use plasma torches, cutting discs and whatever tools you deem necessary to get the job done. Also, as with the wall conversion, remember to wear protective gear because those cut pieces will be sharp and can hurt you if you're not careful. Be sure to smooth out all rough edges and fill any gaps in the metal walls with a sealant to ensure that your container house is airtight and will not allow pests to enter. Next is creating the frames for the doors and windows, and after that, you install the doors and windows and hang them to the frames.

Making the Openings

The process is pretty much the same as cutting through the container walls to make more space to make the openings for walls and frames. Like before, take accurate measurements of the required opening and mark it on the container wall. As we mentioned earlier, you can use a cardboard model of the window (don't forget to include the frame), which will help you get exact measurements. Then, cut through the walls with a torch or other tools.

A plasma cutter is probably the best tool to use here because it gives the cleanest lines and the steel you cut can be reused, unlike other tools that might damage that spare steel. If you can't get your hands on one or

CHAPTER 15
Exterior

Installation Base for The Exterior Part

Next, before we can wrap up the construction process, you will also have to finish the shipping containers' exterior. You have a lot of options here, like with most of the preceding phases. One thing that will make a difference in this process is whether you have added exterior insulation to the shipping container. With insulation or not, you can work on giving the exterior of your home a beautiful finish that will impress visitors.

With Insulation

Having exterior insulation is always a good idea, as we have mentioned so often throughout the book. Fortunately, you have some super options to finish the container's external walls if you have already added spray foam. You can paint over the insulation or add stucco to cover it. The important thing is to keep the insulation sealed and covered, so it could be saved from direct sunlight and external conditions, which might compromise its integrity and cause its insulation to fail.

Painting

Painting is a great approach to finish your shipping container's external walls, but you must use specific kinds of paint. When painting over spray foam, you should use water-based acrylic or latex paint and avoid oil-based paint because it can damage the insulation. Also, steer clear from high-gloss paint since it will make any unevenness on the container pop's surface, unlike semigloss or flat paint, which can hide such inconsistencies and cover the insulation properly.

Before you start painting the container, inspect all the exterior walls of the shipping containers you will paint and look for rough edges or uneven parts. If you encounter those, use sandpaper to even them out so the paint will look smooth and uninterrupted on the exterior. Remember to always wear protective gear like goggles and masks to avoid inhaling fumes or having any particles enter your eyes.

After sanding the rough parts, you can start painting the exterior surfaces of the shipping container. You need at least three coats of paint for the exterior surfaces, and you can choose the tool you prefer to disperse the paint, whether it is a spray gun, brush or roller. However, for exteriors, spray guns would be the best choice, as they provide consistent paint coverage and are also the fastest to get the job done. If you use one, be sure to test it on cardboard first to see how powerful the flow is. Rollers are also good, but are slower than paint guns, while brushes are the slowest, but give excellent control over the paint.

Some experts recommend applying a wax sealant over the paint once you're done painting. This will give the exterior surfaces of the shipping container a much better finish. Use several thin layers rather than one or two thick ones. Let each layer dry off first before applying the new one.

Stucco

Stuccoing the shipping container's exterior surfaces looks quite good and is one option to consider. However, if you're going to apply stucco, it is best that the spray foam has rough edges and finish so the stucco can latch on more easily. (This is one more thing to consider when making a decision on outside insulation.) A distinct advantage about stucco is that it is easy to deal with; you'll just purchase mixes that only need water added before being applied. Ask the salesperson about the surface area that can be covered by each bag of stucco mix.

Before you get to work, make sure you cover the ground around your shipping container home with a plastic sheet to protect it from any spills. To apply stucco, you first need to use adhesive to fix beading to the corners so that the beading is straight with no inconsistencies. Mix the stucco powder with water in a bucket and leave it for five minutes or so. Before you add the stucco, you need to make sure the external insulation surface is wet so that the material can grip onto the walls, using a hose to water it down and keep it moist.

Then, using a steel trowel, work your way from the bottom of the shipping container wall to the top, applying the stucco as you go. Ensure that the stucco is applied evenly across the surface, using long strokes instead of short ones. The process of adding stucco needs to be done within half an hour of mixing, so make sure you are working within that time frame. In terms of layers, follow the same approach as you would with paint; apply several thin ones rather than a couple of thick ones. When the layer is still wet, rake each one after applying the stucco, which will provide a better grip for the next layer. For a more elegant finish to the stucco, you can finish the last layer with a polystyrene float.

Without Insulation

While exterior insulation is important and can help keep a lot of heat and cold out of your shipping container home, you can still skip it and focus on interior insulation. The plus side is your exterior will look like a shipping container home, which looks cool and adds to the uniqueness of the design and shows the true origins of the place you're living in.

Painting

If you're going to leave the exterior of your shipping container home exposed, the least you should do is paint it. An exterior without insulation looks good and shows the hard work you did to make this a home, but it also leaves the exterior exposed to elements. So, a layer of latex-based paint here can make a world of difference. The exterior would still look like a shipping container, but you will have provided a layer that can provide protection against rust and leakage.

Latex-based paint can improve the longevity of your shipping container home and keep the exterior looking nice, too. Before applying the paint, clean the surface of the shipping container, remove any stickers, and use sandpaper and grinders to get rid of the rust that might be lingering on the exterior walls. Using a blade for this chore could prove tricky – and also dangerous. Remember to also cover the ground with plastic sheets before painting.

Experts recommend using alkyd enamel paint for the exterior of a shipping container, and you can apply it using brushes, roller, or spray guns. Like before, the spray gun is the fastest option with the most consistent cover, but you can also use the other two. Use also at least three layers of paint here on the exterior walls.

CHAPTER 16
Other Ideas

Improving a building property can be costly. These changes are, therefore, necessary to make properties more desirable and valuable. Fortunately, homeowners may still use certain products that can be converted for other purposes, and a 20-foot shipping container is one of the best things to invest in.

There are certainly other container sizes. However, due to its size and price, a 20-foot container is suitable for houses. Moreover, homeowners may build various structures with these containers. Any of the following are below.

Garage

One of the best housing structures that can be constructed with a 20-foot cargo container is a garage. Naturally, people who buy vehicles must take care of their investment. That is why it is important to have your garage. This structure helps you shield your vehicles from various wind and rain problems, which can rust and dent your cars.

Shed/Workshop

People with shipping containers can build this structure. Many people spend their time outside on their lawns. However, it can be impossible to spend time in backyards during hot days.

Fortunately, people may use containers to build sheds. Individuals can minimize construction tasks by using shipping containers. These containers can easily withstand unstable weather conditions.

Green House

If you like growing plants and flowers, people can also use containers to build greenhouses. People can choose from different styles of containers, and containers can easily be turned into greenhouses. You can easily grow different plant and flower species in your backyard by getting a greenhouse.

Storage Unit

Another great thing a homeowner can build a storage unit with a 20-foot shipping container. Naturally, almost all homeowners want to upgrade their houses. The easiest way to upgrade homes is to buy new appliances. These may cause unforeseen accidents because of inadequate living space. By building container storage units, homeowners may provide a secure location for their goods.

Swimming Pool

People can use shipping containers to create a swimming pool. Building a pool is one of the major investments made by homeowners to develop their properties. Homeowners must invest in various services from building to earthmoving services during the construction of pools.

You must also spend a large sum of money to complete your project. By using shipping containers, homeowners can quickly and effectively reduce their expenditures and activities. These are just some of the marvelous uses of shipping containers.

Considerations of Containers: Floors

The wooden floors of shipping containers have been a fiercely debated topic in recent years. As retired containers are becoming more creative, they proclaim a positive environmental impact on their recycling efforts.

Many container floors on the other side of the equation are produced using exotic hardwood trees. Although still a renewable resource, one of those trees that has been cut down to produce container floors takes 50 to 60 years to grow.

The chemicals applied in the container floors during

processing are another thing to consider. Wood preservatives, which contain various organochlorine pesticides such as aldrin, dieldrin, chlordane, and lindane, are approved in Australia for treating lumber used in cargo containers as structural parts.

As a result, manufacturers process all containers according to Australian standards; they have agreed it is hard to withdraw the containers from the fleet for any country and face potential penalties and sanctions if an unapproved container has been caught in Australia.

Analysis of these floors and applied insecticides on the container floors has been calculated. The physical accumulation of insecticide from the floor surface is the primary source of toxins. In items recently handled in laminated sawn timber, the highest levels of insecticide residues were found.

Toxins like these dissipate significantly after many years so that the chemicals can be isolated with special preparation, cleaning, and sealing using epoxy finishes. Experts believe that there is practically no danger if a barrier inhibits the exhaust gasification of substances. This is analogous to earlier corrected paint hazards in older buildings.

If the data plate is still in the container, the chemicals initially applied to the floor should be known. Naturally, the data plate cannot support you if the floor has been damaged and modified anywhere along the line. Nor will you learn what was shipped and what leaked during its high seas voyage.

If a container for permanent housing, such as a section of container home, is used, the desired course of action is to remove the original flooring, dispose of it properly, and add new floors. You are looking at ten plywood sheets in a 40-point container plus labor; you can include those costs in your budget if you intend to live or serve food out of a container.

CHAPTER 17

FAQ

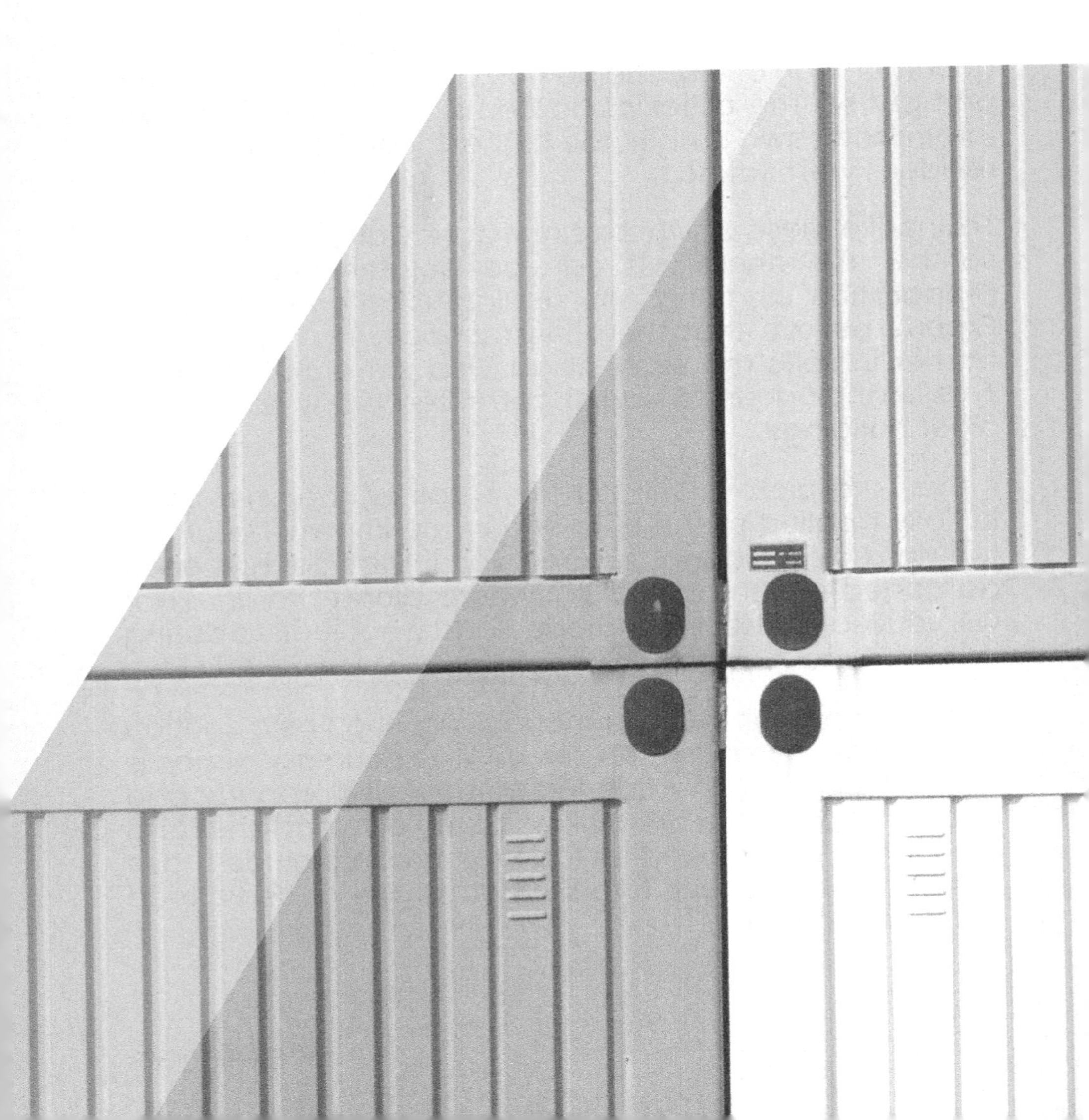

As with any subject in the world, there are some questions that are constantly being asked. They may be questions that you have or questions that other people ask you that you are looking for simple, straightforward answers to. Regardless of why you are looking for them, this is going to answer some of the most frequently asked questions about shipping container houses and their construction. Many of these questions are covered in more depth in the previous pages of this book, but I hope this can serve as your quick reference guide.

Will Not a Shipping Container Become a Hot Oven When It Is Used as A Home or Office?

Not at all. Realistically, any structure base, whether wood, concrete, block or steel panels, has the potential to become a hot oven. All structures require insulation and ventilation. As long as you make sure your shipping container has adequate ventilation and is well insulated you won't have any problems with temperature regulation.

Why Am I Having Such a Hard Time Buying a Shipping Container?

I see many large piles in many places around me.

It is a common thought that if you see a stack of shipping containers, they must be surplus. This is not always true. Shipping companies tend to keep containers in stock, so they are available for export shipments at a moment's notice. This means that even if you see what appears to be a surplus of containers, the company may not have any containers for sale. As a general rule, most companies only sell older containers to the public.

Isn't A Shipping Container Home Going to Rust and Corrode?

Keep in mind that a shipping container is built to transport goods across the ocean. This means they are built for moisture and salt water. A shipping container is finished with a distinct non-corrosive steel that is coated with a ceramic coating that makes it virtually rust-proof. These precautions also prevent mildew and are highly antiseptic.

Why Would Anyone Want to Live in A Home That Appears Similar to A Shipping Container?

Many of the houses that are built with shipping containers look nothing like shipping containers when they are finished. The shipping container is often just the shell. It can be covered with stucco, wood, vinyl siding or anything else. In this regard, not everyone chooses to cover the shipping container. Some people like the look of the container and choose to keep all or part of the shipping container in its natural state.

What Cities Would Ever Permit Shipping Container Homes to Invade Their Pristine Neighborhoods?

There have been very few instances where a planning commission wouldn't allow a shipping container home into their community. Since there are ways to make the exterior appear the same as all the houses surrounding the shipping container home, there isn't any real reason to deny the request. In any case, where the initial request was denied, proper education from the ISBU organization was enough to overturn the decision and make shipping container homes willingly acknowledged in cities and towns all over the world.

Will A Shipping Container Home Obey with Building Codes?

The building code that people are most concerned about is height. Most building codes require an eight-foot ceiling. A standard shipping container is eight feet six inches tall, which means it meets the requirement. Some of the shipping container models are even taller than the standard container.

Are Shipping Containers Earthquake Proof?

A shipping container might roll everywhere a little if it were to take a direct hit in an earthquake. However, it will not collapse and, depending on the interior, would keep the occupants safe from harm. A shipping container home is at least one hundred times safer in an earthquake than a conservative housing assembly, making it a great choice for those who live in areas where earthquakes and other natural disasters are commonplace.

Are Container Homes Sustainable?

Some of the reasons why shipping container homes cannot be regarded to be sustainable are:

- Premature recycling: Used containers often get recycled too soon as there are various homes that are built with only single-use containers.

- Remnant toxins: The majority of the shipping containers get treated with toxic paints along with toxic chemicals for the flooring to maintain durability. Also, there are containers that carried toxic cargo.

- Inefficient recycling: One shipping container, if only recycled as steel, can yield plenty of metal studs for building fourteen houses of the exact same size in comparison to container homes.

- COR-TEN steel: There are shipping containers made from COR-TEN steel, which has lots of environmental baggage, along with corrosion and rust.

Since most of the shipping container houses that can be found are made from used containers, there are professionals in the industry who see this as a type of upcycling. But this form of upcycling is quite inefficient, as long as the above reasons.

Can Shipping Container Homes Turn Out to Be Toxic?

The sole purpose of shipping containers is to transport goods for as many trips as they can. In order to improve their durability, containers are often treated with toxic chemicals. They are not really manufactured for being used as living spaces. Also, the containers might get used for the transportation of toxic goods. The paint used on the walls is very tough and contains loads of chemical compounds that someone would not want in their house. It is always suggested to sand down the existing paints, down to the layer of the raw steel. You can also add extra layers on the walls and floors for sealing the toxic materials.

Do I Need to Get a Foundation for My Shipping Container Home?

Yes, you will need a proper foundation for your container home for two reasons.

- You cannot allow steel to touch the ground as the moisture from the ground might lead to rusting.

- The shipping container needs to be placed on a stable and non-shifting base.

CONCLUSION

These rusty old things we call shipping containers have quite a few benefits for an unsuspecting and needy world. There have been several grass roots projects to use shipping containers as alternative housing in poverty-stricken countries and they have also been used as temporary shelters for victims of natural disasters. There are many ways that the standard shipping container can be used to make the world a better place. Here are a few of the best examples!

- **Shipping Container Housing for the Poor**

Container homes are a great way to provide affordable housing for the less fortunate. Shipping container homes have been placed in economically deprived areas all over the world to put a roof over the financially poorer among us. All you have to do is have a crane to put one down throw on some insulation, install basic electricity and plumbing and you have yourself a great low-income housing unit. The trend is now catching on worldwide and even in the poorest of regions; whether in the economically deprived American South or South Africa these sturdy shipping containers of yore are providing a great place to live for those who really need it.

- **Shipping Container Dorms for Students**

College students are usually a little bit strapped for cash with all of their tuition expenses and the price of books. This usually relegates them to overcrowded college dormitories they can't afford or to living in a rundown house in a rough, crime-infested part of town with 10 other roommates to split the bill! But in many parts of the world, universities have jumped on the bandwagon of the shipping container and have created whole student apartment complexes.

Simply by stacking several containers on top of each other, they have produced instant student housing! Some of these professionally designed student apartments have rent as low as $200 a month! This is unheard of at most other college campuses. Some even come complete with high-speed internet! Get ready to benefit university students; because these sleek shipping container dorms are the wave of the future!

- **Shipping Container Hurricane Relief Shelters**

With the insane number of hurricanes that much of North America has been subjected to in recent times, hurricane relief shelters have been a much-needed resource. After the U.S. was pummeled by brutal storms such as Harvey, Irma, and Maria in rapid succession, emergency relief organizations such as FEMA were running out of places to put the survivors. It was out of this necessity that many had the idea of simply dropping shipping containers off in affected regions such as Texas, Florida, and Puerto Rico. These shelters are ready-made, and with the most minor of modifications, they can be made into a perfect relief shelter for those who need it so desperately.

- **Shipping Container Military Barracks**

The military has actually been using shipping containers as temporary barracks for a while now. Predating the commercial sector's use of shipping containers as homes by several decades, they were even used during the first Iraq War in 1991. These freighters can happily withstand a barrage of weapons fire, so dropping these guys down on an enemy beachhead for troops to seek shelter in is of great advantage.

- **Shipping Containers to Scale Back Your Footprint**

Everybody talks about their environmental footprint nowadays and how they can reduce it. But a shipping container is a means of reducing your impact on this planet almost entirely! You can live in one of these metal hulks and not put any extra strain on the ecosystem whatsoever. You are not using any harmful materials in it its construction, and if you opt for solar, you don't even have to plug into the grid! Another reason that shipping conditions are so good for the planet is the fact that you are basically recycling sturdy building material that would otherwise have ended up in a landfill. So yes, use these shipping containers to scale back that footprint of yours!

Your Ship Has Arrived!

As strange as it may sound, shipping containers could very well be life-changing in their scope. The idea that you can use something such as a shipping container, which has been sent all over the world to ports far and wide, can be used for permanent housing is nothing short of remarkable. These freighters are built perfectly for the job and with a very minimal amount of effort and finances, just about anyone can be a proud new shipping container home owner! Keep your eyes peeled, folks, and keep looking to the horizon, because your ship has arrived!

BONUS CHAPTER

StepByStep Process for Construction

Having a shipping container house obviously has some incredible benefits. It's something that a lot of people are interested in and it's obviously the reason you're reading this book in the first place.

However, there comes a crucial point where you must decide if you would like to undertake the building process yourself or use contractors to perform the build.

Companies that will complete the entire process for you and others will simply work with you along the way.

The decision ultimately depends on your confidence in your construction and management skills and your budget. To undertake construction of this type, you must have a certain amount of construction skill and knowledge. You must be able to get the building permits for your area, acquire the necessary tools and have the ability to follow all the steps of the construction process.

If you don't feel comfortable doing everything yourself, you can hire some contractors to lend a hand with certain aspects (such as window placement), or you can hire a company to do it all for you. The result will be faster and often higher quality construction if you're not a professional builder, but it will usually cost more and you won't get the same satisfaction as if you completed the project on your own.

If you decide to build it yourself, you need to follow the following steps.

Step 1: Design

At this stage, you will start by brainstorming the basic ideas of your container home, its size, and what it should include.

How many rooms would you like? How big should it be? What features should it include? This is where you answer all of these questions. You can then start drawing out sample ideas or base your designs on pre-existing plans.

A great tool to use is SketchUp which allows you to draw in 3D. It's commonly used for designing regular homes, but you can also obviously use it for shipping container homes to get a better idea of how the finished product will look.

Once you have the plans thought out, you should go to an architect or builder to help you finalize the plans and have them drawn up by a professional. The architect will be able to tell you which of your plans are structurally possible and will often have great suggestions for various improvements. Once you have your official plans drawn up, it's on to the next stage.

Step 2: Lay the Foundation

You will need a solid foundation to build your container home on, just like any regular home. Unless you are a professional in this field, it's best to hire a contractor for this job. You need to make sure that the foundation is done correctly, with areas set out to fit the plumbing.

If you're planning a poured concrete foundation, it's a good idea to embed steel fittings into the concrete where the corners of the shipping container will rest. This will help support the container and consult with an engineer or architect about the best choice of foundation for your particular design.

Step 3: Purchase the Container/s

The next step will be the purchase of the containers themselves and their delivery on site. You will need to choose between new or used containers, and you will also need to choose a container size that fits your plan.

Next, if your plan requires any extreme modifications to your container, such as the removal of a large segment of the container, you will need to do so. If not, then it is time to place your shipping container on the foundation. It is best to do this with a crane, which will obviously be one of the big construction expenses. In some cases, a forklift can also be used to maneuver the container.

At this point, the container is basically an empty shell, sitting on a solid foundation. Now it's time to get down to the real work of turning that empty container into a fully functional home.

Step 4: Connect the Containers

Most likely, your house will consist of more than one container. At this point, they will simply be sitting side by side on their foundations, but will not be structurally joined.

The easiest way for DIYers to join these containers together is with a series of bolts drilled through the containers. This is not the most solid way to join the containers together, but it is quick, inexpensive and allows you to uncouple them in the future if you want to make modifications. A more solid way to join the containers would be to weld them together. If you connect the containers with bolts, you can always weld them together later to make the construction more structurally sound.

Step 5: Add Reinforcements

Before removing any walls, we need to make sure that the building is structurally sound. Working with an engineer or architect is essential for this. In addition, you will not get building approval if you do not have the necessary reinforcements.

The reinforcements required vary from one house to another and depend on the municipality's regulations. Typically it will consist of a number of steel beams to help support the structure.

Unless you're a structural engineer, you should seek professional advice in this area. Otherwise, the whole build will be compromised.

Step 6: Add a Roof

Depending on your plan, a roof may not even be necessary. However, it can look great on a shipping container house and aid in proper water runoff. This is useful if you plan on installing a water tank as part of your design, as your roof can be made to collect water.

It is easier to add the roof before doing any trimming while everything is still stable and connected. Be careful when designing your roof and decide if you want to add insulation or not. This can be a great way to add insulation to your container home before construction actually gets underway. You'll be able to live comfortably in one part of the house while others are being finished.

The simplest is a shed-type roof, which allows for rain runoff on both sides of the structure. Insulation can be easily installed, and the roof can be tile or simply sheet steel or tin.

Step 7: Cutouts

Now that you have your shell, it's time to cut out the openings for doorways and windows. Consult your plan, and mark out where the cuts should be made.

You can make these cuts with a cutting torch, a grinder, or a plasma cutter. If you want a clean finish and are worried about charring the remaining walls and flooring, it's best to hire a welder for this job.

Also, make sure to add holes necessary to fit any plumbing or gas fixtures.

Step 8: Add the Flooring

Now you need to add whatever flooring you'd like for your container.

Some people like to completely remove the existing soil in the containers, while others prefer to simply cover it. The choice is yours and obviously depends on the particular container you have purchased and the plans you are following.

A great choice for container homes is floating floorboards. They look like regular floorboards but are fitted much more simply, almost like tiles. They're simple to look after and give your container a real homely feel. But obviously, the choice is up to your personal preferences.

Step 9: Seal Cracks

There will be openings and cracks where the containers join that you have cut doorways. You'll need to seal these up and make sure the joints are clean.

One way to do this would be to weld them together once more. A professional welder best does this to ensure a flush finish.

At this point, the great thing is to add foam insulation between the walls where the gaps are. Fill the gap with foam insulation; either weld the joins shut afterward, or use a foam sealant. Then you can want to add a backer rod and caulk to create a nice finish.

Step 10: Framing

It's now time to add wooden frames to the areas that you've cut. These are the areas where you'll be adding doorways and windows. If you don't have framing experience, it may be best to hire a builder.

You will be creating the frames and attaching them to the container. You can attach them through a combination of bolts and screws. Make sure that you use treated screws to ensure that they last and aren't corrupted.

Step 11: Add Doors and Windows

Now that you've got all of your framing done, it's now a simple job of adding your doors and windows. Most people can purchase a pre-made door and fit it quite simply. Windows, however, can be quite a difficult skill. You will probably need to get the glass custom cut to fit your plan, and this requires a glazier. Then they will fit the windows for you.

Step 12: Interior Framing & Adding Dry Wall

Now that everything is structurally sound, sealed, and with doors and windows fitted, it's time to finish up the inside of your home. This framing should be quite simple to install and is there to hold the drywall in place.

Step 13: Wire Your Container Home

This is now the easiest point to wire your container home. This step does require an electrician. For obvious reasons, DO NOT try and do this yourself unless you're a licensed professional.

Add the wiring to all of the areas you need, and then it's on to the next step.

Step 14: Insulate

You may have already added some insulation between the walls before you sealed them, and you may have added some to the roof. It's a good idea to add additional insulation to your walls now, and then you may end up hanging drywall.

You can also choose to insulate the outside of your container house. It all depends on the look you want to give it. If the exterior of your home is covered with some cedar or vinyl finish, you can easily insulate underneath this finish.

Some people, however, want their container home to look like a container on the outside and display the steel wall of the original container. The choice is yours.

Step 15: Finishing Touches

Now your container home is pretty much complete. It's time to add the finishing touches. Get the final fittings added, such as toilets, sinks, air-conditioners, fans, and light-fittings. Once this is done, it's time to add furniture and move in–congratulations!

Made in the USA
Coppell, TX
11 March 2022